In Memoriam
Lars Inge Hedberg
1935–2005

Memoirs
of the
American Mathematical Society

Number 882

An Axiomatic Approach
to Function Spaces,
Spectral Synthesis, and
Luzin Approximation

Lars Inge Hedberg
Yuri Netrusov

July 2007 • Volume 188 • Number 882 (third of 4 numbers) • ISSN 0065-9266

American Mathematical Society
Providence, Rhode Island

2000 *Mathematics Subject Classification.*
Primary 46E35; Secondary 26B35, 31B15, 31C15, 31C45, 41A17.

Library of Congress Cataloging-in-Publication Data

Hedberg, Lars Inge, 1935-
An axiomatic approach to function spaces, spectral synthesis, and Luzin approximation /Lars Inge Hedberg, Yuri Netrusov.
  p. cm. — (Memoirs of the American Mathematical Society, ISSN 0065-9266 ; no. 882)
"July 2007, volume 188, number 882 (third of 4 numbers)."
Includes bibliographical references.
ISBN 978-0-8218-3983-6 (alk. paper)
  1. Function spaces. 2. Spectral synthesis (Mathematics). 3. Approximation theory. I. Netrusov, Yuri. II. Title.
QA323.H45 2007
515'.73—dc22   2007060756

Memoirs of the American Mathematical Society

This journal is devoted entirely to research in pure and applied mathematics.

Subscription information. The 2007 subscription begins with volume 185 and consists of six mailings, each containing one or more numbers. Subscription prices for 2007 are US$649 list, US$519 institutional member. A late charge of 10% of the subscription price will be imposed on orders received from nonmembers after January 1 of the subscription year. Subscribers outside the United States and India must pay a postage surcharge of US$38; subscribers in India must pay a postage surcharge of US$43. Expedited delivery to destinations in North America US$53; elsewhere US$130. Each number may be ordered separately; *please specify number* when ordering an individual number. For prices and titles of recently released numbers, see the New Publications sections of the *Notices of the American Mathematical Society*.

Back number information. For back issues see the *AMS Catalog of Publications*.

Subscriptions and orders should be addressed to the American Mathematical Society, P. O. Box 845904, Boston, MA 02284-5904, USA. *All orders must be accompanied by payment.* Other correspondence should be addressed to 201 Charles Street, Providence, RI 02904-2294, USA.

Copying and reprinting. Individual readers of this publication, and nonprofit libraries acting for them, are permitted to make fair use of the material, such as to copy a chapter for use in teaching or research. Permission is granted to quote brief passages from this publication in reviews, provided the customary acknowledgment of the source is given.

Republication, systematic copying, or multiple reproduction of any material in this publication is permitted only under license from the American Mathematical Society. Requests for such permission should be addressed to the Acquisitions Department, American Mathematical Society, 201 Charles Street, Providence, Rhode Island 02904-2294, USA. Requests can also be made by e-mail to reprint-permission@ams.org.

Memoirs of the American Mathematical Society is published bimonthly (each volume consisting usually of more than one number) by the American Mathematical Society at 201 Charles Street, Providence, RI 02904-2294, USA. Periodicals postage paid at Providence, RI. Postmaster: Send address changes to Memoirs, American Mathematical Society, 201 Charles Street, Providence, RI 02904-2294, USA.

© 2007 by the American Mathematical Society. All rights reserved.
Copyright of this publication reverts to the public domain 28 years
after publication. Contact the AMS for copyright status.
This publication is indexed in *Science Citation Index*®, *SciSearch*®, *Research Alert*®,
CompuMath Citation Index®, *Current Contents*®/*Physical, Chemical & Earth Sciences*.
Printed in the United States of America.

∞ The paper used in this book is acid-free and falls within the guidelines
established to ensure permanence and durability.
Visit the AMS home page at http://www.ams.org/

10 9 8 7 6 5 4 3 2 1   12 11 10 09 08 07

Contents

Introduction. Notation	1
Chapter 1. A Class of Function Spaces	5
1.1. Definitions and Basic Properties	5
1.2. Some Lemmas	16
1.3. Proof of Theorem 1.1.14	24
1.4. Some Lemmas on Orthogonalization	29
1.5. Proof of Theorem 1.1.15	33
1.6. Homogeneous Spaces	37
1.7. Proof of Theorem 1.6.12	43
1.8. Proof of Theorem 1.6.11	44
Chapter 2. Differentiability and Spectral Synthesis	47
2.1. Capacities and Differentials	47
2.2. Spectral Synthesis	52
2.3. Spectral Synthesis in Spaces of Distributions	57
2.4. Invariant Subspaces and a Theorem of Whitney	64
Chapter 3. Luzin Type Theorems	67
3.1. Luzin Approximation of Functions	67
3.2. Luzin Approximation of Distributions	77
Appendix. Whitney's Approximation Theorem in $L_p(\mathbf{R}^N)$, $p > 0$	87
Bibliography	95

Abstract

We define axiomatically a large class of function (or distribution) spaces on N-dimensional Euclidean space. The crucial property postulated is the validity of a vector-valued maximal inequality of Fefferman–Stein type. The scales of Besov spaces (B-spaces) and Lizorkin–Triebel spaces (F-spaces), and as a consequence also Sobolev spaces, and Bessel potential spaces, are included as special cases. The main results of Chapter 1 characterize our spaces by means of local approximations, higher differences, and atomic representations. In Chapters 2 and 3 these results are applied to prove pointwise differentiability outside exceptional sets of zero capacity, an approximation property known as spectral synthesis, a generalization of Whitney's ideal theorem, and approximation theorems of Luzin (Lusin) type.

Received by the editor May 15, 2004, in revised form November 16, 2004.

2000 *Mathematics Subject Classification.* Primary 46E35; Secondary 26B35, 31B15, 31C15, 31C45, 41A17.

The first author is grateful for a grant from the ESPRC (GR/N12985/01), and for the hospitality of the University of Sussex.

The second author has been the holder of an ESPRC Advanced Fellowship (GR/A00249/01). He is also grateful for several travel grants from the Swedish Natural Science Research Council (NFR), and for the hospitality of Linköping University.

Introduction. Notation

The Sobolev spaces $W_p^m(\mathbf{R}^N)$ of functions, whose derivatives (understood in the weak or distribution sense) of order not exceeding m belong to the Lebesgue space L_p, $1 \leq p \leq \infty$, are an indispensable tool in analysis. They have been generalized to Bessel potential spaces (fractional order Sobolev spaces) L_p^s, where s can be any real number, Besov spaces (or B-spaces) $B_{p,\theta}^s$, and Lizorkin–Triebel spaces (or F-spaces) $F_{p,\theta}^s$, and the theory of the B- and F-spaces has been extended to allow all $p > 0$ and $\theta > 0$. The B-spaces have many applications, and appear naturally for example as restrictions of Sobolev spaces. The main importance of the F-spaces is perhaps that they include the spaces L_p^s, $1 < p < \infty$, and the Hardy spaces H_p, $0 < p \leq 1$, in the same scale of spaces. The first chapter of H. Triebel's book [**54**] gives a good survey of the development of the theory.

Usually results for B-spaces and F-spaces are given separate proofs. The purpose of this work is first (Chapter 1) to give a unified treatment of the scales of B-spaces and F-spaces by including them in a larger class of function (or distribution) spaces which is defined by simple axioms, and then (Chapters 2 and 3) to extend some theorems of analysis to this general setting.

In Section 1.1 we first define certain classes of sequence spaces E, which will be used throughout the paper. The unifying link, which makes the theory possible, is the assumption that the spaces E satisfy a vector-valued maximal function inequality of Fefferman–Stein type.

Given E we define two spaces of functions or distributions, denoted $YL(E)$ and $Y(E)$, by means of representations as sums $\sum f_i$ of entire functions of exponential type with $\{f_i\} \in E$. After proving some of the basic properties of these spaces we formulate two of the main theorems of the paper, Theorems 1.1.14, and 1.1.15. In these theorems, which generalize known results for B- and F-spaces, we give several equivalent definitions of the spaces $YL(E)$ and $Y(E)$, including characterizations by means of local approximations, higher differences, and atomic representations.

The proofs of Theorem 1.1.14, and of the lemmas required, occupy Sections 1.2 and 1.3, and Theorem 1.1.15 is proved in Sections 1.4 and 1.5.

Sections 1.6 – 1.8, which are not necessary for reading the following sections, are devoted to analogous results for so called homogeneous spaces, corresponding to the homogeneous Besov and Lizorkin–Triebel spaces which are often referred to as $\dot{B}$- and $\dot{F}$-spaces.

In Chapter 2 we apply the results of Chapter 1 to an approximation property known as spectral synthesis, and to an extension of Whitney's ideal theorem. In Section 2.1 we define capacities associated to our spaces and prove a theorem about pointwise L_p-differentiability outside exceptional sets of zero capacity. In Sections 2.2 – 2.4 we give extensions to the present general setting of the results given in Chapter 10 of the book [**4**].

In the case of spaces of Besov and Lizorkin–Triebel type the results of Sections 2.2 and 2.4 were announced by Netrusov in the short note [**35**]. In the special case of L_p^s, $1 < p < \infty$, proofs were published in [**4**], Chapter 10, but proofs of the general results have not been previously available. Section 2.3 is devoted to an extension of the spectral synthesis theorem to Besov spaces of distributions.

Chapter 3 is devoted to approximation theorems of Luzin type for the spaces studied in Chapter 1. In Section 3.1, which is independent of the results of Sections 2.2 – 2.4, we prove a theorem which improves earlier such results even in the case of Sobolev spaces. In part this result was announced by Netrusov in [**32**], but the proof has remained unpublished.

This section is continued in Section 3.2 with an analogous theorem for spaces of distributions, which seems to be a question that has not been previously studied.

Finally, in order to make the paper more self-contained, we give as an appendix a complete proof of the Whitney theorem on polynomial approximation in $L_p(\mathbf{R}^N)$ for $0 < p < \infty$ which plays a crucial role in the theory.

The purpose of this joint paper is to give a detailed and accessible exposition of the results sketched above. Since the theory is developed from the beginning, the reader is not required to be familiar with the theory of B- and F-spaces.

NOTATION. Throughout the paper we will use the following notation:

$\mathbf{Z}$, $\mathbf{R}$ — the integers, the real line.

$\mathbf{Z}^N$ — N-tiples of integers, $k = (k_1, k_2, \ldots, k_N)$, $k_i \in \mathbf{Z}$, $i = 1, 2, \ldots, N$.

$\mathbf{R}^N$ — the real N-dimensional Euclidean space, equipped with the norm $|x| = |(x_1, x_2, \ldots, x_N)| = (x_1^2 + x_2^2 + \ldots + x_N^2)^{1/2}$.

$\mathbf{N} = \{1, 2, \ldots\}$, the natural numbers. $\mathbf{N}_0 = \{0\} \cup \mathbf{N}$.

N — the dimension of the space $\mathbf{R}^N$.

$B(x, a)$ — the open ball in $\mathbf{R}^N$ with radius $a > 0$ and center at x.

$Q_{0,0}$ — the unit cell $[0, 1)^N$ in $\mathbf{R}^N$.

$\{Q_{i,k}\}_{(i,k) \in \mathbf{Z} \times \mathbf{Z}^N}$ — the family of dyadic cells

$$Q_{i,k} = 2^{-i}(k + [0, 1)^N) = 2^{-i}(k + Q_{0,0}), \quad i \in \mathbf{Z}, \quad k \in \mathbf{Z}^N.$$

$Q_{i,k}(a)$, $a > 0$ — the closed cube concentric to $Q_{i,k}$ and with sidelength $a2^{-i}$.

$\chi(A)$ or χ_A — the characteristic function for the set A.

$\chi_{i,k}$ — the characteristic function for $Q_{i,k}$.

D^α with a multi-index $\alpha = (\alpha_1, \alpha_2, \ldots, \alpha_N)$ — a partial differentiation operator of order $|\alpha| = \sum_{i=1}^N \alpha_i$.

T_h and $\Delta_h = T_h - I$ with $h \in \mathbf{R}^N$ — translation and difference operators; $T_h f(x) = f(x + h)$, $\Delta_h f(x) = f(x + h) - f(x)$.

$\Delta_h^m = (T_h - I)^m$ — difference operator of order m.

C^m, $m \in \mathbf{N} \cup \{\infty\}$ — the m times continuously differentiable functions, not necessarily bounded. In situations where we use the norm

$$\|f\|_{C^m(A)} = \max_{|\alpha| \le m} \sup_{x \in A} |D^\alpha f(x)|,$$

it is understood that we are dealing with bounded functions.

C_0^m — the subspace of C^m of compactly supported functions.

$|A| = \int_A dx$, $A \subset \mathbf{R}^N$ — the Lebesgue measure.

$L_p(A)$, $0 < p \le \infty$, $A \subset \mathbf{R}^N$ — the Lebesgue spaces, equipped with the (quasi)-norm defined by $\|f\|_{L_p(A)}^p = \int_A |f(x)|^p \, dx$, modified in the usual way for $p = \infty$.

$L_{p,\text{loc}}$ — the space of measurable functions on $\mathbf{R}^N$, belonging to $L_p(A)$ for all compact $A \subset \mathbf{R}^N$.

$\mathcal{D} = C_0^\infty$, $\mathcal{D}'$, $\mathcal{D}'^m$ — the usual spaces of test functions, distributions, and distributions of order at most m.

$\mathcal{S}$, $\mathcal{S}'$ — the Schwartz spaces of test functions and temperate distributions.

$\mathcal{F}$, $\mathcal{F}^{-1}$ — the Fourier transformation, and its inverse.

$M_{r,d}f$ — the maximal function, defined for measurable functions f, and $r > 0$, $d \geq 0$ by

$$M_{r,d}f(x) = \sup_{a>0}\left(a^{-N}\int_{B(0,a)}|f(x+y)|^r(1+|y|)^{-rd}\,dy\right)^{1/r}.$$

$M_{r,0}f$ is denoted $M_r f$.

The action of a distribution f on a test function φ is denoted $\langle f, \varphi \rangle$.

The convolution of a distribution f and a test function φ is denoted $f * \varphi$ and defined by $f * \varphi(x) = \langle f, \varphi(x - \cdot) \rangle$.

$\mathfrak{P}_m$, $m = -1, 0, 1, 2, \ldots$, is the set of polynomials of degree at most m. (We define $\mathfrak{P}_{-1} = \{0\}$). The notation $f \perp \mathfrak{P}_m$ for a function or distribution f means that $\int_{\mathbf{R}^N} f\pi\,dx = 0$, or $\langle f, \pi \rangle = 0$, for all polynomials $\pi \in \mathfrak{P}_m$.

Let $A \subset \mathbf{R}^N$ be bounded, let $m = 0, 1, 2, \ldots$, and let X be a normed or quasi-normed space of functions defined on $\mathbf{R}^N$. The degree of local approximation is

$$\mathcal{E}_m(f, A, X) = \inf_{\pi \in \mathfrak{P}_{m-1}} \|f - \pi\|_{X(A)},$$

where $X(A)$ is the restriction of X to A.

We adopt the convention that the letter C (without any index) is used to denote a constant, whose exact value is irrelevant and can change even within a chain of inequalities.

CHAPTER 1

A Class of Function Spaces

1.1. Definitions and Basic Properties

Everywhere in the following the letter E denotes a quasi-Banach space of sequences of Lebesgue measurable functions on $\mathbf{R}^N$, which has the lattice property with respect to the natural ordering.

More precisely this means that there is a non-negative function $\|\cdot\|$ on E (a quasi-norm), which has the same properties as a norm, except for the triangle inequality, and in addition satisfies the following conditions:

(i) The metric space $(E, \|\cdot\|)$ is complete.
(ii) If $\{f_i\}_{i=0}^\infty \in E$, and $\{g_i\}_{i=0}^\infty$ is a sequence of measurable function such that $|g_i| \leq |f_i|$ a.e. for all i, it follows that $\{g_i\}_{i=0}^\infty \in E$, and
$$\|\{g_i\}_{i=0}^\infty\|_E \leq \|\{f_i\}_{i=0}^\infty\|_E.$$
(iii) There is a constant $C_E \geq 1$ such that

(1.1.1) $$\|F + G\|_E \leq C_E(\|F\|_E + \|G\|_E)$$

for all $F, G \in E$.

It is a well-known property of quasi-norms that (iii) can be replaced by the following:

(iii′) There exist constants $0 < \kappa \leq 1$, and $C'_E \geq 1$, such that for any family $\{F_i\}_{i=0}^j$ of elements in E one has the inequality

(1.1.2) $$\left\|\sum_{i=0}^j F_i\right\|_E^\kappa \leq C'_E \sum_{i=0}^j \|F_i\|_E^\kappa.$$

Here κ can be chosen so that $0 < \kappa \leq \kappa_0$, where $(2C_E)^{2\kappa_0} = 2$, and then $C'_E = (2C_E)^{2\kappa}$.

The property (iii′) was found, apparently independently, by Tosio Aoki [6] and S. Rolewicz [44]; see also Rolewicz [45], Th. 3.2.1. It is also proved in many other places, see e.g. Bergh–Löfström [7], Lemma 3.10.1, DeVore–Lorentz [15], Ch. 2, Th. 1.1, or J. Heinonen [25], Prop. 14.5. The ranges of κ and C'_E given above come from [25].

The following definition will play a central role in what follows.

DEFINITION 1.1.1. Let $\varepsilon_+, \varepsilon_- \in \mathbf{R}$, $0 < r < \infty$, and $d \geq 0$. We define a class $S(\varepsilon_+, \varepsilon_-, r, d)$ of spaces E by saying that $E \in S(\varepsilon_+, \varepsilon_-, r, d)$ if the following conditions are satisfied:

(a) The linear operators S_+ (left shift) and S_- (right shift), defined by
$$S_+(\{f_i\}_{i=0}^\infty) = \{f_{i+1}\}_{i=0}^\infty,$$
$$S_-(\{f_i\}_{i=0}^\infty) = \{f_{i-1}\}_{i=0}^\infty, \quad f_{-1} = 0,$$

are continuous on E, and their norms satisfy for all $j \in \mathbf{N}$ the inequalities
$$\|(S_+)^j\| \leq C_1 2^{-j\varepsilon_+}, \quad \text{and} \quad \|(S_-)^j\| \leq C_2 2^{j\varepsilon_-},$$
i.e., there are constants C_1 and C_2, independent of j and $\{f_i\}_{i=0}^\infty$, such that
$$\|(S_+)^j(\{f_i\}_{i=0}^\infty)\|_E \leq C_1 2^{-j\varepsilon_+}\|\{f_i\}_{i=0}^\infty\|_E,$$
$$\|(S_-)^j(\{f_i\}_{i=0}^\infty)\|_E \leq C_2 2^{j\varepsilon_-}\|\{f_i\}_{i=0}^\infty\|_E.$$

(b) The operator $\mathcal{M}_{r,d}$, defined by
$$\mathcal{M}_{r,d}(\{f_i\}_{i=0}^\infty) = \{M_{r,d}f_i\}_{i=0}^\infty,$$
is bounded on E, i.e.,
$$\|\{M_{r,d}f_i\}_{i=0}^\infty\|_E \leq C\|\{f_i\}_{i=0}^\infty\|_E.$$
Set $S(\varepsilon_+, \varepsilon_-, r) = \bigcup_{d \geq 0} S(\varepsilon_+, \varepsilon_-, r, d)$.

REMARK. Note that $\varepsilon_+ \leq \varepsilon_-$, otherwise the space is trivial. In fact, for any sequence $F = \{f_i\}_{i=0}^\infty$ in E and all $j \geq 0$
$$F = (S_+)^j \circ (S_-)^j(F),$$
whence
$$\|F\|_E \leq C_1 C_2 2^{-j(\varepsilon_+ - \varepsilon_-)},$$
which can be made arbitrarily small if $\varepsilon_+ - \varepsilon_- > 0$.

EXAMPLE. Let Ω denote the class of positive functions ω on $(0, \infty)$ with the property that there is a constant C such that $\omega(x) \leq C\omega(y)$ whenever $\frac{1}{2} \leq x/y \leq 2$. For $\omega \in \Omega$ we denote by l_θ^ω the quasi-Banach space of sequences $\{a_i\}_{i=0}^\infty$ such that the quasi-norm $(\sum_{i=0}^\infty (|a_i|\omega(2^i))^\theta)^{1/\theta}$ is finite (Banach space if $\theta \geq 1$).

The principal examples of spaces E belonging to $S(\varepsilon_+, \varepsilon_-, r)$ are the spaces $l_\theta^\omega(L_p)$ with $0 < p, \theta \leq \infty$ and $L_p(l_\theta^\omega)$ with $0 < p < \infty$, $0 < \theta \leq \infty$ and $\omega \in \Omega$, and with quasi-norms (norms when $p, \theta \geq 1$)
$$\|\{f_i\}_{i=0}^\infty\|_{l_\theta^\omega(L_p)} = \left\|\{\|f_i\|_{L_p(\mathbf{R}^N)}\}_{i=0}^\infty\right\|_{l_\theta^\omega} = \left(\sum_{i=0}^\infty ((\|f_i\|_{L_p(\mathbf{R}^N)})\omega(2^i))^\theta\right)^{1/\theta},$$
and
$$\|\{f_i\}_{i=0}^\infty\|_{L_p(l_\theta^\omega)} = \left\|\|\{f_i(\cdot)\}_{i=0}^\infty\|_{l_\theta^\omega}\right\|_{L_p(\mathbf{R}^N)} = \left\|\left(\sum_{i=0}^\infty (|f_i(\cdot)|\omega(2^i))^\theta\right)^{1/\theta}\right\|_{L_p(\mathbf{R}^N)}.$$

One can easily check that these spaces satisfy (1.1.2) with $\kappa = \min\{1, p, \theta\}$ and $C'_E = 1$, that ε_+ is any constant ε such that $\omega(t)/t^\varepsilon$ is C-increasing, and that ε_- is any ε such that $t^\varepsilon/\omega(t)$ is C-increasing. (A function $f(t)$ is C-increasing if there is a constant C such that $f(x) \leq Cf(y)$ whenever $x \leq y$.)

If $\omega(t) = t^\lambda$, $-\infty < \lambda < \infty$, we write l_θ^λ for l_θ^ω. Then ε_+ is any $\varepsilon \in (-\infty, \lambda]$, and ε_- is any $\varepsilon \in [\lambda, +\infty)$. Thus, in the most important cases we can choose $\varepsilon_+ = \varepsilon_-$. However, these two numbers play different parts in many of the proofs, so for greater transparency it is useful to keep them notationally different.

It follows from the maximal theorems of Hardy–Littlewood–Wiener (see e.g. E. M. Stein [48]) and of Fefferman–Stein ([17], see also Stein [49], and K. F. Andersen and R. T. John [5]) that r can be any number in $(0, p)$ in the case of $l_\theta^\omega(L_p)$, or in $(0, \min\{p, \theta\})$ in the case of $L_p(l_\theta^\omega)$.

These theorems are basic in the theory of B- and F-spaces, see e.g. the historical remarks in Triebel [53], Section 2.3.5.

DEFINITION 1.1.2. If $E \in S(\varepsilon_+, \varepsilon_-, r)$, we denote by E_{dis} the space of sequences of constants $\{s_{i,k}\}_{(i,k) \in \mathbf{N}_0 \times \mathbf{Z}^N}$ such that $\{f_i\}_{i=0}^\infty \in E$, if the functions f_i are defined by

$$f_i = \sum_{k \in \mathbf{Z}^N} s_{i,k} \chi_{i,k}, \quad i \in \mathbf{N}_0.$$

The space is normed by

$$\|\{s_{i,k}\}_{(i,k) \in \mathbf{N}_0 \times \mathbf{Z}^N}\|_{E_{\text{dis}}} = \|\{f_i\}_{i=0}^\infty\|_E.$$

The following two lemmas are easy consequences of Definition 1.1.1, and will play an important part in the following. The first one illustrates the importance of condition (b) in the definition. The role of the maximal theorem is well formulated by M. Frazier and B. Jawerth in [20], p. 228, and the lemma is essentially Lemma 3.1 in [20]; see also [21].

LEMMA 1.1.3. Let $E \in S(\varepsilon_+, \varepsilon_-, r)$, let $b \geq 1$, and set $\widetilde{\chi}_{i,k} = \chi(Q_{i,k}(b))$. Then there is a constant C such that for all sequences $\{s_{i,k}\} \in E_{\text{dis}}$

$$(1.1.3) \quad \left\|\left\{\sum_{k \in \mathbf{Z}^N} s_{i,k} \widetilde{\chi}_{i,k}\right\}_{i=0}^\infty\right\|_E \leq C \left\|\left\{\sum_{k \in \mathbf{Z}^N} s_{i,k} \chi_{i,k}\right\}_{i=0}^\infty\right\|_E.$$

Moreover, if a sequence $\{t_{i,k}\}_{(i,k) \in \mathbf{N}_0 \times \mathbf{Z}^N}$ is defined by $t_{i,k} = \sum_l s_{i,l}$, where the sum is taken over all l such that $Q_{i,l}(b) \supset Q_{i,k}$, then

$$(1.1.4) \quad \|\{t_{i,k}\}\|_{E_{\text{dis}}} \leq C \|\{s_{i,k}\}\|_{E_{\text{dis}}}.$$

PROOF. By the lattice property of E we can assume that all $s_{i,k}$ are nonnegative. Set $f_i(x) = \sum_{l \in \mathbf{Z}^N} s_{i,l} \chi_{i,l}(x)$ for $i = 0, 1, \ldots$. Let $x \in Q_{i,k}$. Then

$$(1.1.5) \quad \begin{aligned} \sum_{l \in \mathbf{Z}^N} t_{i,l} \chi_{i,l}(x) = t_{i,k} &= \sum_{Q_{i,l}(b) \supset Q_{i,k}} s_{i,l} \leq \sum_{Q_{i,l}(b) \ni x} s_{i,l} \\ &= \sum_{l \in \mathbf{Z}^N} s_{i,l} \widetilde{\chi}_{i,l}(x) \leq \sum_{Q_{i,l}(b) \cap Q_{i,k} \neq \emptyset} s_{i,l} \\ &\leq \sum_{Q_{i,l}(b+2) \supset Q_{i,k}} s_{i,l} \leq \sum_{Q_{i,l} \subset B(x, \rho 2^{-i})} s_{i,l}, \end{aligned}$$

where $\rho = \frac{1}{2}(b+3)\sqrt{N}$. It follows that

$$(1.1.6) \quad \left(\sum_{l \in \mathbf{Z}^N} t_{i,l} \chi_{i,l}(x)\right)^r \leq \left(\sum_{l \in \mathbf{Z}^N} s_{i,l} \widetilde{\chi}_{i,l}(x)\right)^r \leq C_r \sum_{Q_{i,l} \subset B(x, \rho 2^{-i})} s_{i,l}^r$$

$$\leq C_r 2^{iN} \int_{B(x, \rho 2^{-i})} f_i(y)^r \, dy \leq C_r \rho^N M_r f_i(x)^r \leq C M_{r,d} f_i(x)^r,$$

where $C_r = 1$ for $0 < r \leq 1$, and C_r depends on r and the number of nonzero terms in the sum, i.e., on b and N, for $r > 1$. The last inequality is true for all $d \geq 0$, with the constant C in addition depending on d, because of the fact that for $i \geq 0$ the radii of the balls appearing in the integral are bounded by ρ. The lemma follows from (b) in Definition 1.1.1. □

The next lemma shows the role of the constant ε_+. Denote by $\delta_{i,j}$, $i, j \in \mathbf{Z}$, the Kronecker symbol, i.e., $\delta_{i,j} = 0$ if $i \neq j$, $\delta_{i,i} = 1$, so that $\{\delta_{0,j} f_j\}_{j=0}^\infty = \{f_0, 0, 0, \ldots\}$.

LEMMA 1.1.4. *Let $E \in S(\varepsilon_+, \varepsilon_-, r)$, suppose that $F = \{f_i\}_{i=0}^{\infty} \in E$, and set $g = \sum_{i=0}^{\infty} 2^{i\lambda}|f_i|$. Then for any $\lambda < \varepsilon_+$ there is C so that*
$$\|\{\delta_{0,j}g\}_{j=0}^{\infty}\|_E \leq C\|F\|_E.$$
In particular, $g \in L_{r,\mathrm{loc}}$, and there is C so that for all $R > 0$
$$(1+R)^{-(N/r+d)}\|g\|_{L_r(B(0,R))} \leq C\|F\|_E.$$

PROOF. It is sufficient to assume that $f_i \geq 0$. We have
$$\{\delta_{0,j}g\}_{j=0}^{\infty} = \sum_{i=0}^{\infty} 2^{i\lambda}\{\delta_{0,j}f_i\}_{j=0}^{\infty} = \sum_{i=0}^{\infty} 2^{i\lambda}(S_+)^i\{\delta_{i,j}f_i\}_{j=0}^{\infty}.$$
But by Definition 1.1.1(a)
$$\|(S_+)^i\{\delta_{i,j}f_i\}_{j=0}^{\infty}\|_E \leq C2^{-i\varepsilon_+}\|\{\delta_{i,j}f_i\}_{j=0}^{\infty}\|_E \leq C2^{-i\varepsilon_+}\|F\|_E,$$
whence
$$\|\{\delta_{0,j}g\}_{j=0}^{\infty}\|_E^{\kappa} \leq C \sum_{i=0}^{\infty} \|2^{i\lambda}(S_+)^i\{\delta_{i,j}f_i\}_{j=0}^{\infty}\|_E^{\kappa}$$
$$\leq C\|F\|_E^{\kappa} \sum_{i=0}^{\infty} 2^{i\kappa(\lambda-\varepsilon_+)} = C\|F\|_E^{\kappa},$$
where κ is the constant in (1.1.2).

This implies that $g \in L_{r,\mathrm{loc}}$, since otherwise $M_{r,d}g$ would be identically $+\infty$, which is impossible.

Moreover, for all $x \in B(0,R)$ we have
$$\|g\|_{L_r(B(0,R))} \leq C(1+R)^{N/r+d}M_{r,d}g(x),$$
so that by the lattice property of E,
$$\|g\|_{L_r(B(0,R))}\|\{\delta_{0,j}\chi(B(0,R))\}_{j=0}^{\infty}\|_E \leq C(1+R)^{N/r+d}\|\{\delta_{0,j}M_{r,d}g\}_{j=0}^{\infty}\|_E.$$
Dividing by the positive number $\|\{\delta_{0,j}\chi(B(0,1))\}_{j=0}^{\infty}\|_E$, and applying Definition 1.1.1(b), we obtain for $R \geq 1$
$$(1+R)^{-(N/r+d)}\|g\|_{L_r(B(0,R))} \leq C\|\{\delta_{0,j}M_{r,d}g\}_{j=0}^{\infty}\|_E \leq C\|\{\delta_{0,j}g\}_{j=0}^{\infty}\|_E.$$
This proves the lemma. □

We now define the spaces which will be our main object of study. Recall that if a distribution $f \in \mathcal{S}'$ is such that $\mathrm{supp}\,\mathcal{F}f$ is compact, and if the order of $\mathcal{F}f$ is m, then by the Paley–Wiener–Schwartz theorem, f is an entire function of exponential type, whose restriction to $\mathbf{R}^N$ satisfies $|f(x)| \leq C(1+|x|)^m$. See e.g. Hörmander [26], Theorem 7.3.1. It follows that $M_{r,d}f(x)$ is finite for all $x \in \mathbf{R}^N$ for $d \geq m$.

DEFINITION 1.1.5. *Suppose that $\varepsilon_+, \varepsilon_-, r > 0$, and let $E \in S(\varepsilon_+, \varepsilon_-, r)$. The space $YL(E)$ consists of all functions $f \in L_{r,\mathrm{loc}}$, which have a representation*

(1.1.7) $$f = \sum_{i=0}^{\infty} f_i,$$

converging in $L_{r,\mathrm{loc}}$, such that

(1.1.8) $$\|\{f_i\}_{i=0}^{\infty}\|_E < \infty,$$

(1.1.9) $$f_i \in \mathcal{S}', \quad \mathrm{supp}\,\mathcal{F}f_i \subset B(0, 2^{i+1}), \quad i \in \mathbf{N}_0.$$

DEFINITION 1.1.6. Suppose that ε_+, $\varepsilon_- \in \mathbf{R}$, $r > 0$, and let $E \in S(\varepsilon_+, \varepsilon_-, r)$. The space $Y(E)$ consists of all distributions $f \in \mathcal{S}'$, which have a representation (1.1.7), converging in $\mathcal{S}'$, satisfying (1.1.8) and

(1.1.10) $$\operatorname{supp} \mathcal{F} f_0 \subset B(0,2),$$

(1.1.11) $$\operatorname{supp} \mathcal{F} f_i \subset B(0, 2^{i+1}) \setminus B(0, 2^{i-1}), \quad i \in \mathbf{N}.$$

EXAMPLE. If $E = l_\theta^\lambda(L_p)$ with $\lambda \in \mathbf{R}$, $0 < p, \theta \leq \infty$, or $E = L_p(l_\theta^\lambda)$ with $0 < p < \infty$, $0 < \theta \leq \infty$, then Definition 1.1.6 is a classical definition of the Besov spaces $B_{p,\theta}^\lambda$ and Lizorkin–Triebel spaces $F_{p,\theta}^\lambda$, respectively. See e.g. the books by J. Peetre [42] and H. Triebel [53], [54]. The spaces obtained in Definition 1.1.5 are in this case denoted $BL_{p,\theta}^\lambda$ and $FL_{p,\theta}^\lambda$. They have been less studied (in the cases when the two definitions do not give the same spaces; see Proposition 1.1.12 below), but see Netrusov [33], [34].

It is easily seen that if $\|f\|_{YL(E)}$ is defined by

$$\|f\|_{YL(E)} = \inf \|\{f_i\}_{i=0}^\infty\|_E,$$

where the infimum is taken over all representations of f as in Definition 1.1.5, and $\|f\|_{Y(E)}$ is defined analogously, the spaces $YL(E)$ and $Y(E)$ are quasi-normed spaces, and normed spaces if E is.

We shall prove that the spaces $YL(E)$ and $Y(E)$ are complete, i.e., quasi-Banach or Banach spaces.

For the proof we need the following lemma of so called Pólya–Plancherel type, which plays an important role for the whole theory. It is a slight modification of Theorem 1.3.1 in Triebel [53].

LEMMA 1.1.7. *Let $f \in \mathcal{S}'$, $\operatorname{supp} \mathcal{F} f \subset B(0, R)$, $r > 0$, $d \geq 0$, and $M \in \mathbf{N}_0$. Then there is a constant C, independent of R, such that for all multi-indices α with $|\alpha| \leq M$,*

(1.1.12) $$\sup_{z \in \mathbf{R}^N} \frac{R^{-|\alpha|} |D^\alpha f(x+z)|}{(1+R|z|)^{N/r+d}} \leq C \sup_{a > 0} \left(\frac{1}{a^N} \int_{B(0,a)} \frac{|f(x+y)|^r}{(1+R|y|)^{rd}} \, dy \right)^{1/r}.$$

PROOF. For the reader's convenience we give the proof, essentially following Triebel [53]. The idea of the proof is due to Peetre [41]. For $r \geq 1$ a simpler proof using Hölder's inequality is possible.

We first prove that for any multi-index α and any $\lambda > 0$ there is C so that

(1.1.13) $$\sup_{z \in \mathbf{R}^N} \frac{R^{-|\alpha|} |D^\alpha f(x+z)|}{(1+R|z|)^\lambda} \leq C \sup_{z \in \mathbf{R}^N} \frac{|f(x+z)|}{(1+R|z|)^\lambda}.$$

We can assume that $R = 1$, i.e., that $\operatorname{supp} \mathcal{F} f \subset B(0,1)$; otherwise replace $f(x)$ by $R^{-N} f(x/R)$. It is also no restriction to assume $x = 0$.

Let φ be a function in $\mathcal{S}$ such that $\mathcal{F}\varphi(\xi) = 1$ on $B(0,1)$. Then $f = f * \varphi$, and $D^\alpha f = f * D^\alpha \varphi$. For any $\lambda \geq 0$ there is C such that

$$|D^\alpha \varphi(y)| \leq C(1+|y|)^{-\lambda-N-1} \quad \text{for all } y,$$

so for any $z \in \mathbf{R}^N$,
$$\frac{|D^\alpha f(z)|}{(1+|z|)^\lambda} \leq C \int_{\mathbf{R}^N} \frac{|f(y+z)|}{(1+|z|)^\lambda (1+|y|)^{\lambda+N+1}} dy$$
$$\leq C \int_{\mathbf{R}^N} \frac{|f(y+z)|}{(1+|z+y|)^\lambda (1+|y|)^{N+1}} dy \leq C \sup_{y \in \mathbf{R}^N} \frac{|f(y)|}{(1+|y|)^\lambda},$$
which proves (1.1.13). Here we have used the inequality $(1+|z|)(1+|y|) \geq 1+|z+y|$, and the integrability of $(1+|y|)^{-N-1}$.

Now, we use (1.1.13) to prove
$$(1.1.14) \qquad \sup_{z \in \mathbf{R}^N} \frac{|f(z)|}{(1+|z|)^{N/r+d}} \leq C \sup_{a \geq 1} \left(\frac{1}{a^N} \int_{B(0,a)} \frac{|f(y)|^r}{(1+|y|)^{rd}} dy \right)^{1/r}.$$

By the mean value theorem, for any $\delta > 0$,
$$(1.1.15) \qquad |f(z)| \leq \inf_{|y-z| \leq \delta} |f(y)| + \delta \sup_{|y-z| \leq \delta} |\nabla f(y)|.$$

Here, if $\delta \leq 1$,
$$\inf_{|y-z| \leq \delta} |f(y)| \leq C \left(\frac{1}{\delta^N} \int_{|y-z| \leq \delta} |f(y)|^r dy \right)^{1/r}$$
$$\leq C \frac{(1+|z|)^{N/r+d}}{\delta^{N/r}} \left(\frac{1}{(1+|z|)^N} \int_{|y| \leq 1+|z|} \frac{|f(y)|^r}{(1+|y|)^{rd}} dy \right)^{1/r}.$$

Substituting this and (1.1.13) into (1.1.15) we obtain,
$$\sup_{z \in \mathbf{R}^N} \frac{|f(z)|}{(1+|z|)^{N/r+d}} \leq C\delta^{-N/r} \sup_{a \geq 1} \left(\frac{1}{a^N} \int_{B(0,a)} \frac{|f(y)|^r}{(1+|y|)^{rd}} dy \right)^{1/r}$$
$$+ C\delta \sup_{z \in \mathbf{R}^N} \frac{|f(z)|}{(1+|z|)^{N/r+d}}.$$

The inequality (1.1.14) and the lemma follow if δ is chosen small enough. $\square$

We note the following corollary.

COROLLARY 1.1.8. *Under the assumptions of Lemma 1.1.7 there is a constant $C = C(r,d,N,M)$ such that for $0 < b \leq 1/R$*
$$\mathcal{E}_M(f, B(0,b), L_\infty) \leq C(bR)^M \sup_{a>0} \left(\frac{1}{a^N} \int_{B(0,a)} \frac{|f(y)|^r}{(1+R|y|)^{rd}} dy \right)^{1/r}.$$

PROPOSITION 1.1.9. *Let E satisfy the conditions in Definition 1.1.5. Then the space $YL(E)$ is continuously embedded in $L_{r,\mathrm{loc}}$ and is complete.*

PROOF. The continuity of the embedding follows from Lemma 1.1.4.

In order to prove completeness it suffices to prove that every sum $\sum_{l=1}^\infty f^{(l)}$ of functions $f^{(l)} \in YL(E)$ such that $\sum_{l=1}^\infty \|f^{(l)}\|_{YL(E)}^\kappa < \infty$, is convergent in $YL(E)$. Here κ is the number defined in (1.1.2). By Definition 1.1.5 and the definition of the norm in $YL(E)$, each $f^{(l)}$ has a representation as a sum of entire functions, $f^{(l)} = \sum_{i=0}^\infty f_i^{(l)}$, satisfying (1.1.9), so that
$$\sum_{l=1}^\infty \|\{f_i^{(l)}\}_{i=0}^\infty\|_E^\kappa \leq 2 \sum_{l=1}^\infty \|f^{(l)}\|_{YL(E)}^\kappa < \infty.$$

But E is complete, so $\sum_{l=1}^{\infty}\{f_i^{(l)}\}_{i=0}^{\infty}$ converges to an element $\{f_i\}_{i=0}^{\infty} \in E$. We have to prove that each f_i belongs to $\mathcal{S}'$, and that $\operatorname{supp} \mathcal{F} f_i \subset B(0, 2^{i+1})$.

By Lemma 1.1.7 we have for any x
$$\sup_{|z|\leq 2^{-i}} |f_i^{(l)}(x+z)| \leq C M_{r,d} f_i^{(l)}(x),$$
or equivalently, for any x
$$|f_i^{(l)}(x)| \leq C \inf_{|z|\leq 2^{-i}} M_{r,d} f_i^{(l)}(x+z). \tag{1.1.16}$$

It follows that for any $R \geq 1$
$$\begin{aligned}
2^{-iN/r} \sup_{x \in B(0,R)} |f_i^{(l)}(x)| &\leq C \left(\int_{B(0,R)} M_{r,d} f_i^{(l)}(x)^r \, dx \right)^{1/r} \\
&\leq C(1+R)^d \inf_{x \in B(0,R)} \left(\int_{B(0,2R)} \frac{M_{r,d} f_i^{(l)}(x+z)^r}{(1+|z|)^{rd}} \, dz \right)^{1/r} \\
&\leq C(1+R)^{N/r+d} \inf_{x \in B(0,R)} M_{r,d}(M_{r,d} f_i^{(l)})(x).
\end{aligned} \tag{1.1.17}$$

As in the proof of Lemma 1.1.4
$$(1+R)^{-(N/r+d)} \sup_{x \in B(0,R)} |f_i^{(l)}(x)| \left\| \{\delta_{0,j}\chi(B(0,R))\}_{j=0}^{\infty} \right\|_E$$
$$\leq C 2^{iN/r} \left\| \{\delta_{0,j} M_{r,d}(M_{r,d} f_i^{(l)})\}_{j=0}^{\infty} \right\|_E \leq C 2^{i(N/r-\varepsilon_+)} \left\| \{f_j^{(l)}\}_{j=0}^{\infty} \right\|_E.$$

Since $\left\| \{\delta_{0,j}\chi(B(0,R))\}_{j=0}^{\infty} \right\|_E \geq \left\| \{\delta_{0,j}\chi(B(0,1))\}_{j=0}^{\infty} \right\|_E > 0$ for $R \geq 1$, this proves that
$$\sup_x (1+|x|)^{-(N/r+d)} |f_i^{(l)}(x)| \leq C 2^{i(N/r-\varepsilon_+)} \left\| \{f_j^{(l)}\}_{j=0}^{\infty} \right\|_E, \tag{1.1.18}$$

and consequently
$$\sup_x (1+|x|)^{-(N/r+d)} \sum_{l=1}^{\infty} |f_i^{(l)}(x)| \leq C 2^{i(N/r-\varepsilon_+)} \left(\sum_{l=1}^{\infty} \|f^{(l)}\|_{YL(E)}^{\kappa} \right)^{1/\kappa}.$$

It follows that $\sum_{l=1}^{\infty} f_i^{(l)}(x) = f_i(x)$ with convergence in $\mathcal{S}'$, and uniformly on compact sets. Thus $\operatorname{supp} \mathcal{F} f_i \subset B(0, 2^{i+1})$, $f = \sum_{i=0}^{\infty} f_i \in YL(E)$, and $\sum_{l=1}^{\infty} f^{(l)}$ converges to f in $YL(E)$. $\square$

PROPOSITION 1.1.10. *Let E satisfy the conditions in Definition 1.1.6. Then the space $Y(E)$ is continuously embedded in $\mathcal{S}'$.*

PROOF. Let $\{f_i\}_{i=0}^{\infty} \in E$, and suppose that (1.1.10) and (1.1.11) are satisfied. We claim that $\sum_{i=0}^{\infty} f_i$ converges in the sense of $\mathcal{S}'$.

We first observe that in order to prove that $\sum_{i=0}^{\infty} f_i$ converges in $\mathcal{S}'$, it is sufficient to prove that $\sum_{i=0}^{\infty} 2^{-i\sigma} f_i$ converges in $\mathcal{S}'$ for some $\sigma > 0$. In fact, we shall see that if $\sum_{i=0}^{\infty} f_i$ converges in $\mathcal{S}'$, then $\sum_{i=0}^{\infty} 2^{i\sigma} f_i$ converges in $\mathcal{S}'$ for any $\sigma \in \mathbf{R}$. In order to prove this we choose C_0^{∞} functions φ_i such that $\operatorname{supp} \varphi_0 \subset B(0, 2)$, $\operatorname{supp} \varphi_i \subset B(0, 2^{i+1}) \setminus B(0, 2^{i-1})$, $i = 1, 2, \ldots$, and $\varphi_i(\xi) = 1$ on $\operatorname{supp} \mathcal{F} f_i$. Set $\Phi_1 = \sum_{i=0}^{\infty} 2^{2i\sigma} \varphi_{2i}$ and $\Phi_2 = \sum_{i=0}^{\infty} 2^{(2i+1)\sigma} \varphi_{2i+1}$. Then Φ_1 and Φ_2 belong to C^{∞} and $\Phi_1 \psi, \Phi_2 \psi$ belong to $\mathcal{S}$ for any ψ in $\mathcal{S}$.

Suppose that $\sum_{i=0}^{\infty} f_i$ converges in $\mathcal{S}'$. Then $\sum_{i=0}^{\infty} \mathcal{F} f_i$ also converges in $\mathcal{S}'$, i.e., $\sum_{i=0}^{\infty} \langle \mathcal{F} f_i, \psi \rangle$ converges for any $\psi \in \mathcal{S}$. But then

$$\sum_{i=0}^{\infty} \langle 2^{i\sigma} \mathcal{F} f_i, \psi \rangle = \sum_{i=0}^{\infty} \langle 2^{i\sigma} \varphi_i \mathcal{F} f_i, \psi \rangle = \sum_{i=0}^{\infty} \langle \Phi_1 \mathcal{F} f_{2i}, \psi \rangle + \sum_{i=0}^{\infty} \langle \Phi_2 \mathcal{F} f_{2i+1}, \psi \rangle$$
$$= \sum_{i=0}^{\infty} \langle \mathcal{F} f_{2i}, \Phi_1 \psi \rangle + \sum_{i=0}^{\infty} \langle \mathcal{F} f_{2i+1}, \Phi_2 \psi \rangle.$$

Here the last two series are convergent, which proves the convergence in $\mathcal{S}'$ of $\sum_{i=0}^{\infty} 2^{i\sigma} \mathcal{F} f_i$, and hence of $\sum_{i=0}^{\infty} 2^{i\sigma} f_i$.

We claim that $\sum_{i=0}^{\infty} 2^{-i\sigma} f_i$ converges in $\mathcal{S}'$ for $\sigma > N/r - \varepsilon_+$. As in (1.1.18)

(1.1.19) $$\sup_x (1+|x|)^{-(N/r+d)} |f_i(x)| \le C 2^{i(N/r - \varepsilon_+)} \|\{f_j\}_{j=0}^{\infty}\|_E.$$

We set $\sigma - N/r + \varepsilon_+ = \iota$, so that $\iota > 0$. Then

$$\sum_{i=0}^{\infty} 2^{-i\sigma} \sup_x (1+|x|)^{-(N/r+d)} |f_i(x)| \le \sum_{i=0}^{\infty} 2^{-i\iota} \|\{f_j\}_{j=0}^{\infty}\|_E.$$

This proves that

$$(1+|x|)^{-(N/r+d)} \sum_{i=0}^{\infty} 2^{-i\sigma} f_i(x)$$

converges uniformly in $\mathbf{R}^N$. The claim, and the continuity of the embedding follow immediately. $\square$

PROPOSITION 1.1.11. *Let E satisfy the conditions in Definition 1.1.6. Then the space $Y(E)$ is complete.*

PROOF. The proof is the same as that of Proposition 1.1.9. $\square$

The elements of $Y(E)$ are in general distributions, and not functions, and if $0 < r < 1$ the functions in $YL(E)$ are not in general distributions. However, the following is true.

PROPOSITION 1.1.12. *Suppose that $\varepsilon_+, \varepsilon_- \in \mathbf{R}$, $r > 0$, and let $E \in S(\varepsilon_+, \varepsilon_-, r)$. Then $YL(E) = Y(E)$ if $\varepsilon_+ > N \max\{\frac{1}{r} - 1, 0\}$.*

PROOF. Let $\varepsilon_+ > N \max\{\frac{1}{r} - 1, 0\}$, and suppose that $\{f_i\}_{i=0}^{\infty} \in E$ satisfies (1.1.9). We claim that for $R \ge 1$

(1.1.20) $$(1+R)^{-(N/r+d)} \sum_{i=0}^{\infty} \int_{B(0,R)} |f_i(y)|\, dy \le C \|\{f_i\}_{i=0}^{\infty}\|_E.$$

If $r \ge 1$, we have $\varepsilon_+ > 0$, and (1.1.20) follows from Lemma 1.1.4.

Now let $0 < r < 1$ and $\varepsilon_+ - N/r + N = \iota > 0$. By (1.1.17) for $R \geq 1$

$$2^{-iN(1-r)/r} \int_{B(0,R)} |f_i(x)|\, dx$$

$$\leq 2^{-iN(1-r)/r} \sup_{x \in B(0,R)} |f_i(x)|^{1-r} \int_{B(0,R)} |f_i(x)|^r\, dx$$

$$\leq C(1+R)^{(N/r+d)(1-r)} \inf_{x \in B(0,R)} M_{r,d}(M_{r,d}f_i)(x)^{1-r} \int_{B(0,R)} |f_i(x)|^r\, dx$$

$$\leq C(1+R)^{N/r+d} \inf_{x \in B(0,R)} M_{r,d}(M_{r,d}f_i)(x).$$

Thus for each $R \geq 1$ and each $x \in B(0,R)$

$$R^{-(N/r+d)} \sum_{i=0}^{\infty} \int_{B(0,R)} |f_i(y)|\, dy \leq \sum_{i=0}^{\infty} 2^{iN(1-r)/r} M_{r,d}(M_{r,d}f_i)(x).$$

It follows as before that

$$R^{-(N/r+d)} \sum_{i=0}^{\infty} \int_{B(0,R)} |f_i(y)|\, dy \leq C \left(\sum_{i=0}^{\infty} 2^{-i\kappa\iota} \right)^{1/\kappa} \|\{f_i\}_{i=0}^{\infty}\|_E,$$

which is (1.1.20).

Now let $f \in Y(E)$, and suppose that $f = \sum_{i=0}^{\infty} f_i$ is a representation satisfying (1.1.8), (1.1.10) and (1.1.11), and converging in $\mathcal{S}'$. By (1.1.20) the series converges in $L_{1,\text{loc}}$ ($L_{r,\text{loc}}$ if $r > 1$ by Lemma 1.1.4), and consequently the distribution f can be identified with a function in $L_{1,\text{loc}}$. It follows that $f \in YL(E)$, and $Y(E) \subset YL(E)$.

To prove the converse inclusion, let $f \in YL(E)$, and suppose that $f = \sum_{i=0}^{\infty} f_i$ is a representation satisfying (1.1.8) and (1.1.9), and converging in $L_{r,\text{loc}}$. It follows from (1.1.20) that the series converges to f in $\mathcal{S}'$.

We have to prove that f has a representation $f = \sum_{i=0}^{\infty} g_i$ satisfying (1.1.10) and (1.1.11), and such that $\|\{g_j\}_{j=0}^{\infty}\|_E \leq C\|\{f_j\}_{j=0}^{\infty}\|_E$. Let $\psi_0 \in \mathcal{S}$ be such that $\operatorname{supp} \mathcal{F}\psi_0 \subset B(0,2)$ and $\mathcal{F}\psi_0(\xi) = 1$ for $\xi \in B(0,1)$. Set

$$\psi_j(x) = 2^{(j+1)N} \psi_0(2^{j+1}x) - 2^{jN} \psi_0(2^j x)$$

for $j \in \mathbf{N}$, so that $\mathcal{F}\psi_j(\xi) = \psi_0(2^{-j-1}\xi) - \psi_0(2^{-j}\xi)$, $\sum_{j=0}^{i} \mathcal{F}\psi_j(\xi) = 1$ on $B(0,2^{i+1})$, and $\operatorname{supp}\mathcal{F}\psi_j \subset B(0,2^{j+2}) \setminus B(0,2^j)$. Then

$$f_i = f_i * \left(\sum_{j=0}^{i} \psi_j \right) = \sum_{j=0}^{i} f_i * \psi_j.$$

This gives, if the change in order of summation is justified,

$$(1.1.21) \qquad f = \sum_{i=0}^{\infty} f_i = \sum_{i=0}^{\infty} \sum_{j=0}^{i} f_i * \psi_j = \sum_{j=0}^{\infty} \sum_{i=j}^{\infty} f_i * \psi_j = \sum_{j=0}^{\infty} g_j$$

with $g_j = \sum_{i=j}^{\infty} f_i * \psi_j$. But $\sum_{i=j}^{\infty} f_i$ converges in $\mathcal{S}'$, so $g_j = \left(\sum_{i=j}^{\infty} f_i \right) * \psi_j$, and hence (1.1.10) and (1.1.11) are satisfied.

We can write $\{g_j\}_{j=0}^{\infty} = \{\sum_{i=0}^{\infty} f_{i+j} * \psi_j\}_{j=0}^{\infty} = \sum_{i=0}^{\infty} \{f_{i+j} * \psi_j\}_{j=0}^{\infty}$. Let φ be a function in $\mathcal{S}$ such that $\mathcal{F}\varphi(\xi) = 1$ on $B(0,2^{i+j+1})$. Then $f_{i+j} * \psi_j = f_{i+j} * (\psi_j * \varphi)$, and Lemma 1.1.7 shows that $|f_{i+j} * \psi_j(x)| \leq CM_{r,d}f_{i+j}(x)$. Thus

$$\|\{g_j\}_{j=0}^\infty\|_E^\kappa \leq C \sum_{i=0}^\infty \|\{M_{r,d}f_{i+j}\}_{j=0}^\infty\|_E^\kappa$$
$$= C \sum_{i=0}^\infty \|(S_+)^i(\{f_j\}_{j=0}^\infty)\|_E^\kappa \leq C \sum_{i=0}^\infty 2^{-i\varepsilon_+} \|\{f_j\}_{j=0}^\infty\|_E^\kappa,$$

as claimed. By Lemma 1.1.4 the sums in (1.1.21) are almost everywhere absolutely convergent, which justifies the operations, and proves the inclusion $YL(E) \subset Y(E)$. □

PROPOSITION 1.1.13. *Suppose that ε_+, $\varepsilon_- \in \mathbf{R}$, $r > 0$, and let $E \in S(\varepsilon_+, \varepsilon_-, r)$. Then $Y(E) \subset \mathcal{D}'^L$, i.e., the elements of $Y(E)$ are distributions of order at most L, if $L > N \max\{\frac{1}{r} - 1, 0\} - \varepsilon_+$.*

PROOF. It is proved as in Proposition 1.1.12 that $\sum_{i=0}^\infty 2^{-iL} f_i \in L^1_{\text{loc}}$. The result follows easily. □

In the following two theorems we formulate the basic results for the spaces $YL(E)$ and $Y(E)$, respectively. Note that by Proposition 1.1.12 for a given $E \in S(\varepsilon_+, \varepsilon_-, r)$ these spaces coincide if $\varepsilon_+ > N \max\{\frac{1}{r} - 1, 0\}$.

THEOREM 1.1.14. *Let ε_+, ε_-, $r > 0$, and let $E \in S(\varepsilon_+, \varepsilon_-, r)$. Then, for any p such that $0 < p \leq \infty$ and $\frac{N}{p} > \frac{N}{r} - \varepsilon_+$, and any integer M such that $M > \varepsilon_-$, the following conditions on a function $f \in L_{r,\text{loc}}$ are equivalent (with the usual modification if $p = \infty$):*

(i) $f \in YL(E)$.

(ii) $f \in L_{p,\text{loc}}$, *and the functions*

$$(1.1.22) \qquad g_0(x) = \|f\|_{L_p(B(x,1))},$$

$$(1.1.23) \qquad g_i(x) = 2^{iN/p} \left(\int_{B(x,2^{-i})} |\Delta_z^M f(x)|^p \, dz \right)^{1/p} \quad \text{for } i \in \mathbf{N},$$

satisfy

$$(1.1.24) \qquad \|\{g_i\}_{i=0}^\infty\|_E < \infty.$$

(iii) $f \in L_{p,\text{loc}}$, *and (1.1.24) is satisfied with g_0 defined by (1.1.22) and*

$$g_i(x) = 2^{iN/p} \mathcal{E}_M(f, B(x, 2^{-i}), L_p) \quad \text{for } i \in \mathbf{N}.$$

(iv) *f has a representation*

$$(1.1.25) \qquad f = \sum_{(i,k) \in \mathbf{N}_0 \times \mathbf{Z}^N} s_{i,k} a_{i,k},$$

converging in $L_{r,\text{loc}}$, where the functions $a_{i,k}(2^{-i}(x+k)) = b_{i,k}(x)$ satisfy

$$(1.1.26) \qquad b_{i,k} \in C_0^M([-1,2]^N), \quad \text{and} \quad \|b_{i,k}\|_{C^M} \leq 1,$$

and the coefficients $s_{i,k}$ are such that (1.1.24) is satisfied with

$$(1.1.27) \qquad g_i(x) = \sum_{k \in \mathbf{Z}^N} |s_{i,k}| \chi_{i,k}(x), \quad i \in \mathbf{N}_0.$$

Moreover, the convergence in (1.1.25) holds absolutely almost everywhere, and $g = \sum_{i=0}^\infty g_i$ belongs to $L_{p,\text{loc}}$.

(v) *There are polynomials $\pi_{i,k} \in \mathfrak{P}_{M-1}$, $(i,k) \in \mathbf{N} \times \mathbf{Z}^N$, such that* (1.1.24) *holds with*

(1.1.28) $$g_0 = f, \quad g_i = \sum_{k \in \mathbf{Z}^N} |f - \pi_{i,k}| \chi_{Q_{i,k}(3)} \quad \text{for } i \in \mathbf{N}.$$

Furthermore, in each of the cases (ii), (iii), (iv), *and* (v), (1.1.24) *defines a norm equivalent to* $\|f\|_{YL(E)}$.

The proof of this theorem depends on a number of lemmas, which are given in Section 1.2. The proof itself is given in Section 1.3.

THEOREM 1.1.15. *Let ε_+, $\varepsilon_- \in \mathbf{R}$, $r > 0$, $d \geq 0$ and let $E \in S(\varepsilon_+, \varepsilon_-, r, d)$. Then, for any nonnegative integers L and M such that $L > N\max\{\frac{1}{r} - 1, 0\} - \varepsilon_+$ and $M > \varepsilon_-$, the following conditions on a distribution $f \in \mathcal{S}'$ are equivalent:*
(i) $f \in Y(E)$.
(ii) *The estimate*

(1.1.29) $$\|\{h_i\}_{i=0}^\infty\|_E < \infty$$

is satisfied with

$$h_0(x) = \sup |\langle f, \varphi(\,\cdot\, - x) \rangle|,$$

where the supremum is taken over all $\varphi \in \mathcal{S}$, normalized so that for a fixed number $\lambda > N\max\{\frac{1}{r}, 1\} + d$,

(1.1.30) $$\max_x (1 + |x|)^\lambda |D^\alpha \varphi(x)| \leq 1 \quad \text{for all } \alpha \text{ with } |\alpha| \leq L,$$

and with

$$h_i(x) = 2^{iN} \sup |\langle f, \psi(2^i(\,\cdot\, - x)) \rangle| \quad \text{for } i \in \mathbf{N},$$

where the supremum is taken over all $\psi \in \mathcal{S}$, satisfying (1.1.30), *and such that, in addition, $\psi \perp \mathfrak{P}_{M-1}$.*
(ii') *The estimate* (1.1.29) *is satisfied with*

$$h_0(x) = \sup |\langle f, \varphi(\,\cdot\, - x) \rangle|,$$

where the supremum is taken over all $\varphi \in C_0^\infty(B(0,1))$, such that $\|\varphi\|_{C^L} \leq 1$, and with

$$h_i(x) = 2^{iN} \sup |\langle f, \psi(2^i(\,\cdot\, - x)) \rangle| \quad \text{for } i \in \mathbf{N},$$

where the supremum is taken over all $\psi \in C_0^\infty(B(0,1))$, such that $\|\psi\|_{C^L} \leq 1$, and such that, in addition, $\psi \perp \mathfrak{P}_{M-1}$.
(iii) *f has a representation* (1.1.25), *converging in $\mathcal{S}'$, where the functions $a_{i,k}$ in addition to* (1.1.26) *satisfy $a_{i,k} \perp \mathfrak{P}_{L-1}$ for $i \in \mathbf{N}$, and the coefficients $s_{i,k}$ are such that* (1.1.24) *is satisfied with g_i defined by* (1.1.27).

REMARK. Note that the conditions (ii) and (ii') are analogous to condition (iii) in Theorem 1.1.14. In fact, the functions h_i in the former conditions measure the degree of local polynomial approximation to the distribution f. We return to this point in Section 3.2; see in particular Lemma 3.2.1.

The proof of this theorem again depends on a number of lemmas, mainly in order to achieve the orthogonality conditions. These are given in Section 1.4, and then the proof of the theorem is given in Section 1.5.

REMARK. In the case of B- and F-spaces the results of the last two theorems are essentially known. The definitions 1.1.5 and 1.1.6, which we have chosen as our starting point, go back to J. Peetre [39], [42]. The extension of the theory to quasi-normed spaces ($0 < p < 1$) is also due to Peetre [40], [41]. The characterizations by means of local approximation have their starting point in the work of Yu. A. Brudnyĭ [12]. The representation by atoms is due to M. Frazier and B. Jawerth [19], [20], [21], see also [22], and, independently, to Netrusov [30], [31], [33], [34]. The spaces of type BL and FL were introduced in [33], see also [34]. The books by H. Triebel [53], [54] contain a wealth of results, historical information, and references to earlier work, so there is no need to go further into details here. However, condition (v) in Theorem 1.1.14 does not seem to be previously recorded in the literature. In the case of B-spaces it is easily seen to be equivalent to (iii). (Added 1 September 2005: See also Triebel's recent paper [55].)

1.2. Some Lemmas

In this section we collect a number of lemmas which are required for the proof of Theorem 1.1.14.

LEMMA 1.2.1. *Let $f \in L_{p,\text{loc}}$, $0 < p \leq \infty$, and $M \in \mathbf{N}$. Then there is a constant $C = C(M,p,N)$, such that for any cube Q with side a there is a polynomial $\pi \in \mathfrak{P}_{M-1}$ satisfying (with the usual modification if $p = \infty$):*

$$(1.2.1) \quad |f(x) - \pi(x)| \leq C\left(a^{-N} \int_{B(0,a)} |\Delta_z^M f(x)|^p \, dz\right)^{1/p}$$
$$+ C\left(a^{-2N} \int_{B(0,a)} \int_{Q(2)} |\Delta_z^M f(y)|^p \, dy \, dz\right)^{1/p}$$

for all $x \in Q$.

Moreover, the polynomial π can be chosen independently of p.

PROOF. In order to prove (1.2.1) we start from the following known inequality, which is of a type that goes back to H. Whitney [58], [59], and has been much studied in approximation theory.

Let $0 < p \leq \infty$, let $f \in L_{p,\text{loc}}(\mathbf{R}^N)$, and let $M \in \mathbf{N}$, and $0 < \rho \leq 1$. Then there is a constant $C = C(M,p,\rho,N)$, such that for any cube Q with side a there is a polynomial $\pi \in \mathfrak{P}_{M-1}$ satisfying

$$(1.2.2) \quad \int_Q |f(y) - \pi(y)|^p \, dy \leq C a^{-N} \int_{|z|<\rho a} \int_Q |\Delta_z^M f(y)|^p \, dy \, dz.$$

Since there does not seem to be any easily accessible reference for this result in its full generality, we give a proof as an appendix. See Theorem A.1.

We choose a $\pi \in \mathfrak{P}_{M-1}$ so that (1.2.2) is satisfied with $\rho = \frac{1}{2}$ for the cube $Q(2)$, and set $f - \pi = g$. We have

$$\Delta_z^M g(x) = (T_z - I)^M g(x) = (-1)^M g(x) + \sum_{k=1}^M (-1)^{M-k} \binom{M}{k} g(x+kz).$$

Observing that $\Delta_z^M g(x) = \Delta_z^M f(x)$, we obtain

$$|g(x)| \leq |\Delta_z^M f(x)| + C \sum_{k=1}^{M} |g(x+kz)|,$$

and thus, for $x \in Q$,

$$|g(x)|^p = Ca^{-N} \int_{B(0,\frac{a}{2M})} |g(x)|^p \, dz$$

$$\leq Ca^{-N} \int_{B(0,\frac{a}{2M})} |\Delta_z^M f(x)|^p \, dz + Ca^{-N} \sum_{k=1}^{M} \int_{B(0,\frac{a}{2M})} |g(x+kz)|^p \, dz.$$

By applying (1.2.2) to the terms on the right hand side we find

$$\int_{B(0,\frac{a}{2M})} |g(x+kz)|^p \, dz \leq \int_{Q(2)} |g(y)|^p \, dy \leq Ca^{-N} \int_{|z|<a} \int_{Q(2)} |\Delta_z^M f(y)|^p \, dy \, dz,$$

and (1.2.1) follows.

We now prove the last statement of the lemma. For a function f on a set $A \subset \mathbf{R}^N$ we denote by f^* its non-increasing rearrangement, i.e.,

$$f^*(t) = \inf\{\, s : |E_s| \leq t \,\}$$

for $t \geq 0$. Here $E_s = \{\, x \in A : |f(x)| > s \,\}$, and $|E_s|$ denotes the N-dimensional measure. Clearly

$$f^*(t) \leq t^{-1/p} \|f\|_{L_p(A)} \quad \text{for all } p > 0.$$

There is also a constant $C = C(M, N)$, such that for any polynomial $\pi \in \mathfrak{P}_{M-1}$, restricted to the N-dimensional unit ball $B(0,1) = B$,

$$\|\pi\|_{L_\infty(B)} \leq C \pi^*(\tfrac{1}{2}|B|).$$

In fact, suppose the contrary. Then there are polynomials $\pi_n \in \mathfrak{P}_{M-1}$, $n = 1$, 2, ..., such that $\|\pi_n\|_{L_\infty(B)} = 1$, but $\lim_{n \to \infty} \pi_n^*(\tfrac{1}{2}|B|) = 0$. This sequence of polynomials has a subsequence, which converges uniformly on B to a polynomial $\pi \in \mathfrak{P}_{M-1}$. It follows that $\pi(x) = 0$ on a set of positive measure, which gives a contradiction.

Now, for a given $f \in L_p(B)$, $p > 0$, we choose $\pi_0 \in \mathfrak{P}_{M-1}|_B$ so that

$$(f - \pi_0)^*(\tfrac{1}{4}|B|) \leq 2 \inf_{\pi \in \mathfrak{P}_{M-1}} (f - \pi)^*(\tfrac{1}{4}|B|) \leq 2 \cdot 4^{1/p} |B|^{-1/p} \inf_{\pi \in \mathfrak{P}_{M-1}} \|f - \pi\|_{L_p(B)},$$

and $\pi_p \in \mathfrak{P}_{M-1}|_B$ so that

$$\|f - \pi_p\|_{L_p(B)} \leq 2 \inf_{\pi \in \mathfrak{P}_{M-1}} \|f - \pi\|_{L_p(B)}.$$

Then

$$(\pi_0 - \pi_p)^*(\tfrac{1}{2}|B|) \leq (\pi_0 - f)^*(\tfrac{1}{4}|B|) + (f - \pi_p)^*(\tfrac{1}{4}|B|)$$
$$\leq C \inf_{\pi \in \mathfrak{P}_{M-1}} \|f - \pi\|_{L_p(B)},$$

and hence

$$\|\pi_0 - \pi_p\|_{L_\infty(B)} \leq C \inf_{\pi \in \mathfrak{P}_{M-1}} \|f - \pi\|_{L_p(B)},$$

from which it follows that

$$\|\pi_0 - f\|_{L_p(B)} \leq C \inf_{\pi \in \mathfrak{P}_{M-1}} \|f - \pi\|_{L_p(B)}.$$

$\square$

The following lemma is simple but useful.

LEMMA 1.2.2. *Let $A, B \subset Q$, where Q is a cube with side a. Let $M \in \mathbf{N}$, and $0 < q \leq p \leq \infty$, and suppose that $|A \cap B| \geq \delta|Q| > 0$. Let $f \in L_{p,\mathrm{loc}}$ and suppose that π_A, π_B are polynomials in $\mathfrak{P}_{M-1}$ such that*

$$\|f - \pi_A\|_{L_q(A)} \leq 2\mathcal{E}_M(f, A, L_q) \; ;$$
$$\|f - \pi_B\|_{L_p(B)} \leq 2\mathcal{E}_M(f, B, L_p) \; .$$

Then there is a constant C, independent of a, such that

$$\max_{|\alpha|=m} |D^\alpha(\pi_A - \pi_B)| \leq Ca^{-N/p-m}\mathcal{E}_M(f, Q, L_p), \quad m = 0, 1, 2, \ldots, M-1.$$

PROOF. We have

$$a^{-N/q}\|\pi_A - \pi_B\|_{L_q(A \cap B)}$$
$$\leq Ca^{-N/q}(\|f - \pi_A\|_{L_q(A \cap B)} + \|f - \pi_B\|_{L_q(A \cap B)})$$
$$\leq Ca^{-N/q}\|f - \pi_A\|_{L_q(A \cap B)} + Ca^{-N/p}\|f - \pi_B\|_{L_p(A \cap B)}$$
$$\leq Ca^{-N/q}\mathcal{E}_M(f, A, L_q) + Ca^{-N/p}\mathcal{E}_M(f, B, L_p)$$
$$\leq Ca^{-N/p}\mathcal{E}_M(f, Q, L_p).$$

The lemma now follows from the equivalence of norms on a finite dimensional normed or quasi-normed space, and a scaling argument. □

LEMMA 1.2.3. *Let $0 < q \leq p < \infty$, $f \in L_{p,\mathrm{loc}}$, and $M \in \mathbf{N}$. Let $\pi_{i,k}$, $(i,k) \in \mathbf{N}_0 \times \mathbf{Z}^N$, be polynomials in $\mathfrak{P}_{M-1}$ such that*

$$\|f - \pi_{i,k}\|_{L_q(Q_{i,k}(3))} \leq 2\mathcal{E}_M(f, Q_{i,k}(3), L_q),$$

and let $\varphi_{i,k}$ be functions such that $\mathrm{supp}\,\varphi_{i,k} \subset Q_{i,k}(3)$, $\sum_{k \in \mathbf{Z}^N} \varphi_{i,k}(x) \equiv 1$, and $\|\varphi_{i,k}\|_{L_\infty} \leq 1$ for all $i \in \mathbf{N}_0$. Then the functions $\{f_i\}_{i \in \mathbf{N}_0}$, defined by

$$f_i(x) = \sum_{k \in \mathbf{Z}^N} \pi_{i,k}(x)\varphi_{i,k}(x),$$

converge to the function $f(x)$ almost everywhere and in $L_{p,\mathrm{loc}}$ (in L_p if $f \in L_p$).

PROOF. We first prove convergence almost everywhere. Cf. Theorem 4.8.1 in [**4**]. We fix an x, and we can assume, without loss of generality, that $f(x) = 0$, and that x is a Lebesgue point in the L^q-sense for f, and thus

$$\lim_{i \to \infty, \, Q_{i,k} \ni x} 2^{iN} \int_{Q_{i,k}(5)} |f(y)|^q \, dy = 0.$$

For any i there is a k such that $x \in Q_{i,k}$. Then $\varphi_{i,l}(x) \neq 0$ only if l is such that $Q_{i,k} \subset Q_{i,l}(3)$. But for such l

$$\int_{Q_{i,k}} |\pi_{i,l}|^q \, dy \leq C\int_{Q_{i,k}} |f - \pi_{i,l}|^q \, dy + C\int_{Q_{i,k}} |f|^q \, dy$$
$$\leq C\mathcal{E}_M(f, Q_{i,l}(3), L_q)^q + C\int_{Q_{i,k}} |f|^q \, dy$$
$$\leq C\mathcal{E}_M(f, Q_{i,k}(5), L_q)^q + C\int_{Q_{i,k}} |f|^q \, dy.$$

Trivially
$$2^{iN}\mathcal{E}_M(f, Q_{i,k}(5), L_q)^q \leq 2^{iN}\int_{Q_{i,k}(5)}|f|^q\,dy \to 0\,, \quad \text{as } i \to \infty\,,$$
so by Lemma 1.2.2
$$\max_{y\in Q_{i,k}}|\pi_{i,l}(y)| \leq C\left(2^{iN}\int_{Q_{i,k}}|\pi_{i,l}|^q\,dy\right)^{1/q} \to 0\,, \quad \text{as } i \to \infty\,,$$
and thus $\lim_{i\to\infty} f_i(x) = 0$, which proves almost everywhere convergence.

The convergence in $L_{p,\text{loc}}$ or L_p for $q \leq p < \infty$ is a consequence of the Whitney inequality (1.2.2). This inequality, combined with Lemma 1.2.2, gives for any $Q_{i,k}$
$$\int_{Q_{i,k}(3)}|f(y) - \pi_{i,k}(y)|^p\,dy \leq C2^{iN}\int_{|z|\leq 2^{-i}}\int_{Q_{i,k}(3)}|\Delta_z^M f(y)|^p\,dy\,dz.$$
After multiplication with a suitable compactly supported cut-off function we can assume that $f \in L_p$. Then, for any $i \in \mathbf{N}_0$,
$$\int_{\mathbf{R}^N}|f - f_i|^p\,dy \leq \int_{\mathbf{R}^N}\left(\sum_{k\in\mathbf{Z}^N}|\varphi_{i,k}||f - \pi_{i,k}|\right)^p\,dx$$
$$\leq C\sum_{l\in\mathbf{Z}^N}\int_{Q_{i,l}}\sum_{Q_{i,k}\subset Q_{i,l}(3)}|f - \pi_{i,k}|^p\,dy \leq C\sum_{k\in\mathbf{Z}^N}\int_{Q_{i,k}(3)}|f - \pi_{i,k}|^p\,dy$$
$$\leq C2^{iN}\sum_{k\in\mathbf{Z}^N}\int_{|z|\leq 2^{-i}}\int_{Q_{i,k}(3)}|\Delta_z^M f(y)|^p\,dy\,dz$$
$$\leq C2^{iN}\int_{|z|\leq 2^{-i}}\int_{\mathbf{R}^N}|\Delta_z^M f(y)|^p\,dy\,dz \leq C\max_{|z|\leq 2^{-i}}\int_{\mathbf{R}^N}|\Delta_z^M f(y)|^p\,dy,$$
which approaches zero, as $i \to \infty$. In fact, this last statement is true for continuous f, and continuous functions are dense in L_p for $0 < p < \infty$. $\square$

LEMMA 1.2.4. *Let $\varepsilon_+, \varepsilon_-, r > 0$, $d \geq 0$, $E \in S(\varepsilon_+, \varepsilon_-, r, d)$, $b > 0$, $0 < p \leq \infty$, and $\frac{N}{p} > \frac{N}{r} - \varepsilon_+$. Define an operator T_p on E_{dis} by*
$$T_p(\{s_{i,k}\}_{(i,k)\in\mathbf{N}_0\times\mathbf{Z}^N}) = \{t_{i,k}\}_{(i,k)\in\mathbf{N}_0\times\mathbf{Z}^N},$$
where
$$(1.2.3) \qquad t_{i,k} = 2^{iN/p}\left\|\sum_{j,l}|s_{j,l}|\chi_{j,l}\right\|_{L_p},$$
and the sum is taken over all indices (j,l) such that $Q_{j,l} \subset Q_{i,k}(b)$, and $j \geq i$.

Then T_p maps E_{dis} into E_{dis}, and there is C such that
$$\|T_p(\{s_{i,k}\})\|_{E_{\text{dis}}} \leq C\|\{s_{i,k}\}\|_{E_{\text{dis}}}$$
for all $\{s_{i,k}\} \in E_{\text{dis}}$. Here the constant C depends on the constants appearing in the definition of the class $S(\varepsilon_+, \varepsilon_-, r, d)$, and the constant b.

PROOF. Set
$$\frac{N}{p} - \frac{N}{r} + \varepsilon_+ = 2\iota,$$
so $\iota > 0$. We claim that there is C such that $\|\{t_{i,k}\}\|_{E_{\text{dis}}} \leq C\|\{s_{i,k}\}\|_{E_{\text{dis}}}$. Without loss of generality we can assume all $s_{j,l} \geq 0$. Denote
$$g_j(x) = \sum_{l\in\mathbf{Z}^N} s_{j,l}\chi_{j,l}(x).$$

We have
$$\|\{t_{i,k}\}\|_{E_{\mathrm{dis}}} = \|\{\sum_k t_{i,k}\chi_{i,k}\}_{i=0}^\infty\|_E,$$
where
$$t_{i,k} \le 2^{iN/p}\|\sum_{j=i}^\infty g_j\|_{L_p(Q_{i,k}(b))}.$$

The right hand side is an increasing function of p, so we can without loss of generality assume that $p \ge r$.

We will use the facts that for any sequence $\{a_j\}$ its l_p-norm $(\sum_j |a_j|^p)^{1/p}$ is non-increasing for $0 < p \le \infty$, and that for $0 < p < 1$, and any $\iota > 0$, there is C such that

$$(1.2.4) \qquad \left(\sum_{j=0}^\infty |a_j|^p\right)^{1/p} \le C \sup_{j \ge 0} |a_j| 2^{j\iota} \le C \sum_{j=0}^\infty |a_j| 2^{j\iota}.$$

Then, for $1 \le p < \infty$, by Minkowski's inequality

$$t_{i,k} \le \sum_{j=i}^\infty \left(2^{iN}\int_{Q_{i,k}(b)} g_j^p\, dx\right)^{1/p},$$

whereas for $0 < p < 1$

$$t_{i,k} \le \left(\sum_{j=i}^\infty 2^{iN}\int_{Q_{i,k}(b)} g_j^p\, dx\right)^{1/p}.$$

In either case, since $p \ge r$,

$$2^{iN}\int_{Q_{i,k}(b)} g_j^p\, dx = 2^{iN}\int_{Q_{i,k}(b)} \left(\sum_l s_{j,l}\chi_{j,l}\right)^p dx$$
$$\le \sum_{\{l:Q_{j,l}\cap Q_{i,k}(b)\neq\emptyset\}} s_{j,l}^p 2^{(i-j)N} \le \left(\sum_l s_{j,l}^r 2^{(i-j)N}\right)^{p/r} 2^{(j-i)Np(\frac{1}{r}-\frac{1}{p})}$$
$$\le \left(2^{iN}\int_{Q_{i,k}(b+2)} g_j^r\, dx\right)^{p/r} 2^{(j-i)(\varepsilon_+ -2\iota)}.$$

The modification for $p = \infty$ is easy. Thus, for $1 \le p \le \infty$,

$$t_{i,k}\chi_{i,k}(x) \le \sum_{j=i}^\infty \left(2^{iN}\int_{Q_{i,k}(b+2)} g_j^r\, dy\right)^{1/r} 2^{(j-i)(\varepsilon_+ -2\iota)}$$
$$\le C\sum_{j=i}^\infty M_{r,d}g_j(x) 2^{(j-i)(\varepsilon_+ -2\iota)}.$$

Here the last inequality is true for any $d \ge 0$ because of the fact that the sides of the cubes appearing in the integrals are bounded by $b+2$ for $i \ge 0$.

For $0 < p < 1$, similarly, by (1.2.4)

$$t_{i,k}\chi_{i,k}(x) \leq \left(\sum_{j=i}^{\infty}\left(2^{iN}\int_{Q_{i,k}(b+2)} g_j^r\, dy\right)^{p/r} 2^{p(j-i)(\varepsilon_+ - 2\iota)}\right)^{1/p}$$

$$\leq C\sum_{j=i}^{\infty}\left(2^{iN}\int_{Q_{i,k}(b+2)} g_j^r\, dy\right)^{1/r} 2^{(j-i)(\varepsilon_+ - 2\iota)}2^{(j-i)\iota}$$

$$\leq C\sum_{j=i}^{\infty} M_{r,d}g_j(x) 2^{(j-i)(\varepsilon_+ - \iota)}.$$

We have

$$\left\{\sum_{j=i}^{\infty} M_{r,d}g_j(x) 2^{(j-i)(\varepsilon_+ - \iota)}\right\}_{i=0}^{\infty} = \left\{\sum_{j=0}^{\infty} M_{r,d}g_{i+j}(x) 2^{j(\varepsilon_+ - \iota)}\right\}_{i=0}^{\infty}$$

$$= \sum_{j=0}^{\infty}\{M_{r,d}g_{i+j}(x) 2^{j(\varepsilon_+ - \iota)}\}_{i=0}^{\infty} = \sum_{j=0}^{\infty} 2^{j(\varepsilon_+ - \iota)}(S_+)^j(\{M_{r,d}g_i(x)\}_{i=0}^{\infty}),$$

and thus, by the property (1.1.2) of quasi-norms and Definition 1.1.1,

$$\|\{t_{i,k}\}\|_{E_{\text{dis}}}^{\kappa} = \|\{\sum_k t_{i,k}\chi_{i,k}\}_{i=0}^{\infty}\|_E^{\kappa}$$

(1.2.5)
$$\leq C\sum_{j=0}^{\infty} 2^{\kappa j(\varepsilon_+ - \iota)}\|(S_+)^j(\{M_{r,d}g_i\}_{i=0}^{\infty})\|_E^{\kappa}$$

$$\leq C\|\{M_{r,d}g_i\}_{i=0}^{\infty}\|_E^{\kappa} \sum_{j=0}^{\infty} 2^{-\kappa j\iota}$$

$$\leq C\|\{g_i\}_{i=0}^{\infty}\|_E^{\kappa} = C\|\{s_{i,k}\}\|_{E_{\text{dis}}}^{\kappa},$$

which proves the lemma. $\square$

The lemma contains the following result which we formulate as a corollary. Cf. Lemma 1.1.4.

COROLLARY 1.2.5. *Let ε_+, ε_-, $r > 0$, and let $E \in S(\varepsilon_+, \varepsilon_-, r)$. Let*

$$\{s_{i,k}\}_{(i,k) \in \mathbf{N}_0 \times \mathbf{Z}^N} \in E_{\text{dis}},$$

and set

$$g_i = \sum_{k \in \mathbf{Z}^N} s_{i,k}\chi_{i,k}, \quad i \in \mathbf{N}_0.$$

Then $\sum_{i=0}^{\infty} |g_i| \in L_{p,\text{loc}}$ for any p such that $0 < p \leq \infty$ and $\frac{N}{p} > \frac{N}{r} - \varepsilon_+$, and the series $\sum_{i=0}^{\infty} g_i$ converges almost everywhere, and in $L_{p,\text{loc}}$ (unless $p = \infty$).

PROOF. We assume all $s_{i,k} \geq 0$. By (1.2.5)

$$\|\{\delta_{0,i}t_{0,k}\}_{i,k}\|_{E_{\text{dis}}} \leq \|\{t_{i,k}\}_{i,k}\|_{E_{\text{dis}}} \leq C\|\{s_{i,k}\}_{i,k}\|_{E_{\text{dis}}},$$

and thus by (1.2.3) with $b = 1$ we have for each k

$$t_{0,k} = \left\|\sum_{i,l} s_{i,l}\chi_{i,l}\right\|_{L_p(Q_{0,k})} = \left\|\sum_{i=0}^{\infty} g_i\right\|_{L_p(Q_{0,k})} < \infty,$$

where the first sum is taken over all indices (i,l) such that $Q_{i,l} \subset Q_{0,k}$. $\square$

LEMMA 1.2.6. *Let ε_+, $\varepsilon_- \in \mathbf{R}$, $d \geq 0$, $r > 0$, $E \in S(\varepsilon_+, \varepsilon_-, r, d)$, $b > 0$, and $\lambda > \varepsilon_-$. Define an operator T_λ by*

$$T_\lambda(\{s_{i,k}\}_{(i,k) \in \mathbf{N}_0 \times \mathbf{Z}^N}) = \{t_{i,k}\}_{(i,k) \in \mathbf{N}_0 \times \mathbf{Z}^N},$$

where

(1.2.6) $$t_{i,k} = \sum_{j,l} 2^{-\lambda(i-j)} |s_{j,l}|,$$

the sum being taken over all indices (j,l) such that $j < i$ and $Q_{j,l}(b) \supset Q_{i,k}$.

Then T_λ is continuous from E_{dis} to E_{dis}, and its norm can be estimated from above by means of the constants in the definition of the class $S(\varepsilon_+, \varepsilon_-, r, d)$, and the constant b.

PROOF. It is enough to assume all $s_{j,l} \geq 0$. Denote $\chi(Q_{j,l}(b)) = \widetilde{\chi}_{j,l}$, and set

$$h_j(x) = \sum_{l \in \mathbf{Z}^N} s_{j,l} \widetilde{\chi}_{j,l}(x).$$

Then

$$\{\sum_k t_{i,k} \chi_{i,k}\}_{i=0}^\infty = \{\sum_{j=0}^i 2^{-\lambda(ij)} h_j\}_{i=0}^\infty$$
$$= \{\sum_{j=0}^i 2^{-\lambda j} h_{i-j}\}_{i=0}^\infty = \sum_{j=0}^\infty 2^{-\lambda j}(S_-)^j(\{h_i\}_{i=0}^\infty),$$

so by the quasi-norm property (1.1.2), (a) in Definition 1.1.1, and Lemma 1.1.3

$$\|\{t_{i,k}\}\|_{E_{\mathrm{dis}}}^\kappa = \|\{\sum_k t_{i,k} \chi_{i,k}\}_{i=0}^\infty\|_E^\kappa \leq C \sum_{j=0}^\infty 2^{-\kappa\lambda j} \|(S_-)^j(\{h_i\}_{i=0}^\infty)\|_E^\kappa$$
$$\leq C \|\{h_i\}_{i=0}^\infty)\|_E^\kappa \sum_{j=0}^\infty 2^{\kappa j(\varepsilon_- - \lambda)} \leq C \|\{s_{i,k}\}\|_{E_{\mathrm{dis}}}^\kappa. \qquad \square$$

LEMMA 1.2.7. *Let ε_+, $\varepsilon_- \in \mathbf{R}$, $d \geq 0$, $r > 0$, $E \in S(\varepsilon_+, \varepsilon_-, r, d)$, and $r_1 \leq 1$, $r_1 < r$. Let $\{g_i\}_{i=0}^\infty \in E$, with*

$$g_i(x) = \sum_{k \in \mathbf{Z}^N} s_{i,k} \chi_{i,k}(x),$$

and set

$$T_j g_i(x) = \sum_{k \in \mathbf{Z}^N} |s_{i,k}| \gamma_j(x - k2^{-i}), \quad j \in \mathbf{N}_0,$$

where the functions γ_j are nonnegative and satisfy

$$\gamma_j(x) \leq (1 + |x|2^j)^{-N/r_1} \quad \text{for } |x| \leq 1,$$
$$\gamma_j(x) \leq 2^{-jN/r_1} |x|^{-N/r_1 - d} \quad \text{for } |x| \geq 1.$$

Then there is a constant C, independent of i and j, so that

$$T_j g_i(x) \leq C 2^{(i-j) \max\{N/r, N\}} M_{r,d} g_i(x).$$

PROOF. Cf. Frazier and Jawerth [**19**], Lemma 3.4, and Lemma 4.6.5 in [**4**]. Assume that $s_{i,k} \geq 0$, and set

$$B_0 = \{k \in \mathbf{Z}^N : |k| \leq 2^{1+i-j}\sqrt{N}\}, \quad \text{and}$$
$$B_m = \{k \in \mathbf{Z}^N : 2^{m+i-j}\sqrt{N} < |k| \leq 2^{1+m+i-j}\sqrt{N}\} \quad \text{for } m = 1, 2, \ldots.$$

Let $r \leq 1$. We can without restriction assume that $x \in Q_{i,0}$. Then, for $m \leq j$,

$$\sum_{k \in B_m} s_{i,k} \gamma_j(x - k2^{-i}) \leq \sum_{k \in B_m} s_{i,k}(1 + 2^j|x - k2^{-i}|)^{-N/r_1}$$

$$\leq C2^{-mN/r_1} \sum_{k \in B_m} s_{i,k} \leq C2^{-mN/r_1} \Big(\sum_{k \in B_m} s_{i,k}^r\Big)^{1/r}$$

$$= C2^{-mN/r_1} \Big(2^{iN} \int_{\mathbf{R}^N} \sum_{k \in B_m} s_{i,k}^r \chi_{i,k}(y) \, dy\Big)^{1/r}$$

$$\leq C2^{-mN/r_1} \Big(2^{(m+i-j)N} \rho^{-N} \int_{B(x,\rho)} g_i(y)^r \, dy\Big)^{1/r}$$

$$\leq C2^{-mN(1/r_1 - 1/r)} 2^{(i-j)N/r} M_{r,d} g_i(x),$$

where $\rho = 2^{2+m-j}\sqrt{N}$. Similarly, for $m > j$, with the same ρ,

$$\sum_{k \in B_m} s_{i,k} \gamma_j(x - k2^{-i}) \leq \sum_{k \in B_m} s_{i,k} 2^{-jN/r_1} |x - k2^{-i}|^{-N/r_1 - d}$$

$$\leq C2^{-mN/r_1} 2^{(j-m)d} \sum_{k \in B_m} s_{i,k}$$

$$\leq C2^{-mN/r_1} 2^{(j-m)d} \Big(\sum_{k \in B_m} s_{i,k}^r\Big)^{1/r}$$

$$\leq C2^{-mN/r_1} 2^{(j-m)d} \Big(2^{iN} \int_{\rho/2 < |x| < \rho} g_i(y)^r \, dy\Big)^{1/r}$$

$$\leq C2^{-mN(1/r_1 - 1/r)} 2^{(i-j)N/r} \Big(\rho^{-N} \int_{\rho/2 < |x| < \rho} \frac{g_i(y)^r}{(1 + |x - y|)^{rd}} \, dy\Big)^{1/r}$$

$$\leq C2^{-mN(1/r_1 - 1/r)} 2^{(i-j)N/r} M_{r,d} g_i(x).$$

By assumption $1/r_1 - 1/r > 0$, and it follows by summing over m that $T_j g_i(x) \leq C2^{(i-j)N/r} M_{r,d} g_i(x)$. The same inequality holds true in the case $r > 1$. The proof of this is similar, and is omitted. The lemma follows. □

LEMMA 1.2.8. *Let $\Phi \in C^M$ satisfy*

$$\max_x (1 + |x|)^\lambda |D^\beta \Phi(x)| \leq 1$$

for all $|\beta| \leq M$, and some $\lambda > 0$. Let $\Psi \in C$ satisfy $\Psi \perp \mathfrak{P}_{M-1}$, and

$$\max_x (1 + |x|)^{N+M+\lambda} |\Psi(x)| \leq 1.$$

Set $\Psi_t(x) = t^{-N} \Psi(x/t)$, $t > 0$. Then there is a constant C such that for all x

(1.2.7) $$|\Phi * \Psi_t(x)| \leq C \frac{t^M}{(1 + |x|)^\lambda}, \quad 0 < t \leq 1.$$

PROOF. See Frazier and Jawerth [**19**], Lemma 3.3. For the reader's convenience we give the proof.

By assumption

$$\Phi * \Psi_t(x) = \int_{\mathbf{R}^N} \Psi_t(x - y)\Big(\Phi(y) - \sum_{|\beta| \leq M-1} D^\beta \Phi(x)(y - x)^\beta / \beta!\Big) dy$$

and thus, by Taylor's formula,

$$|\Phi * \Psi_t(x)| \leq \int_{\mathbf{R}^N} |\Psi_t(x-y)||x-y|^M R(x,y)\, dy,$$

where

$$R(x,y) = \max_{|\beta|=M} \max_{0 \leq \theta \leq 1} |D^\beta \Phi(y + \theta(x-y))|/\beta!.$$

For $|x-y| \leq |x|/2$ we have by assumption $R(x,y) \leq C(1+|x|)^{-\lambda}$, and thus

$$\int_{|x-y| \leq |x|/2} |\Psi_t(x-y)||x-y|^M R(x,y)\, dy$$

$$\leq \frac{C}{(1+|x|)^\lambda} \int_{\mathbf{R}^N} |\Psi_t(z)||z|^M\, dz \leq \frac{Ct^M}{(1+|x|)^\lambda}.$$

For $|x-y| \geq |x|/2$ we obtain by the boundedness of R,

$$\int_{|x-y| \geq |x|/2} |\Psi_t(x-y)||x-y|^M R(x,y)\, dy \leq C \int_{|z| \geq |x|/2} |\Psi_t(z)||z|^M\, dz$$

$$\leq Ct^M \int_{|z| \geq |x|/(2t)} |\Psi(z)||z|^M\, dz \leq \frac{Ct^M}{(1+|x|/(2t))^\lambda} \leq \frac{Ct^M}{(1+|x|)^\lambda},$$

since $t \leq 1$. $\square$

The following lemma follows immediately from the previous one by symmetry and a scaling argument.

LEMMA 1.2.9. *Let* $\Phi \in C^M$ *satisfy* $\Phi \perp \mathfrak{P}_{L-1}$, *and*

$$\max_x (1+|x|)^{N+L+\lambda} |D^\beta \Phi(x)| \leq 1$$

for all $|\beta| \leq M$ *and some* $\lambda > 0$. *Let* $\Psi \in C^L$ *satisfy* $\Psi \perp \mathfrak{P}_{M-1}$, *and*

$$\max_x (1+|x|)^{N+M+\lambda} |D^\beta \Psi(x)| \leq 1$$

for all $|\beta| \leq L$. *Then there is a constant* C *such that for all* x *and all* $i, j \in \mathbf{N}_0$

(1.2.8) $$|\Phi_{2^{-i}} * \Psi_{2^{-j}}(x)| \leq C \frac{2^{-(j-i)M}}{(1+2^i|x|)^\lambda}, \quad i \leq j;$$

(1.2.9) $$|\Phi_{2^{-i}} * \Psi_{2^{-j}}(x)| \leq C \frac{2^{-(i-j)L}}{(1+2^j|x|)^\lambda}, \quad i \geq j.$$

1.3. Proof of Theorem 1.1.14

Let $q = \min\{p,r\}$, and denote by (ii*) and (iii*) the conditions (ii) and (iii) with p replaced by q. (Clearly, the latter imply the former.) We will organize the proof in steps according to the following plan. We first prove the equivalence of conditions (i) and (iv), and then we show the chain of implications (ii*) $\Rightarrow$ (v) $\Rightarrow$ (iii*) $\Rightarrow$ (iv). Finally, we prove that (ii) and (iii) follow from (iv).

1.3. PROOF OF THEOREM 1.1.14

Step 1. Let $f \in YL(E)$, and suppose that the representation $f = \sum_0^\infty f_i$ satisfies (1.1.8) and (1.1.9). Let $\omega \in C_0^\infty(Q(3))$ satisfy

(1.3.1) $$\sum_{k \in \mathbf{Z}^N} \omega(x-k) = 1, \quad x \in \mathbf{R}^N.$$

We define constants $s_{i,k}$ and functions $a_{i,k}$ for $i \in \mathbf{N}_0$ and $k \in \mathbf{Z}^N$ by

$$s_{i,k} a_{i,k}(x) = \omega\bigl(2^i(x - k2^{-i})\bigr) f_i(x), \quad \|a_{i,k}(\cdot\, 2^{-i})\|_{C^M(\mathbf{R}^N)} = 1.$$

It follows easily from Lemma 1.1.7 that for $x \in Q_{i,k}$,

$$g_i(x) = |s_{i,k}| \le C M_{r,d} f_i(x).$$

This implies (iv) by (b) in Definition 1.1.1. In particular $\sum_0^\infty f_i$ converges to f in $L_{p,\mathrm{loc}}$ for all p such that $\frac{N}{p} > \frac{N}{r} - \varepsilon_+$ by Corollary 1.2.5.

Step 2. We now prove the implication (iv) $\Rightarrow$ (i). Assume that f has a representation (1.1.25), $f = \sum s_{i,k} a_{i,k}$, converging in $L_{p,\mathrm{loc}}$ for some $p > 0$, and satisfying (1.1.26) with an $M > \varepsilon_-$, (1.1.27), and (1.1.24). By Corollary 1.2.5 the series converges absolutely almost everywhere, and in $L_{p,\mathrm{loc}}$ for $\frac{N}{p} > \frac{N}{r} - \varepsilon_+$, so it is no restriction to assume $p \ge r$. Let

(1.3.2) $$\Psi \in \mathcal{S}, \quad \mathcal{F}\Psi \in C_0^\infty(B(0,2)), \quad \mathcal{F}\Psi(x) = 1 \quad \text{for } |x| \le 1.$$

Set $\Psi_{2^{-j}} = 2^{jN} \Psi(\cdot\, 2^j)$ for $j \in \mathbf{N}_0$, $\psi_0 = \Psi$, and $\psi_j = \Psi_{2^{-j}} - \Psi_{2^{-(j-1)}}$ for $j \in \mathbf{N}$. Then the functions ψ_j, $j \ge 1$, satisfy $\psi_j \perp \mathfrak{P}_{M-1}$, and it follows from the properties of $\mathcal{S}$ and from (1.2.8) in Lemma 1.2.9 that for any $\lambda > 0$ there is C so that

$$|a_{i,k} * \psi_j(x)| \le \frac{C 2^{-(j-i)M}}{(1 + 2^i|x - k2^{-i}|)^\lambda}, \quad 0 \le i \le j, \quad x \in \mathbf{R}^N.$$

Thus, if λ is sufficiently large, $|a_{i,k} * \psi_j(x)| \le C 2^{-(j-i)M} |\gamma_i(x - k2^{-i}|$, where γ_i is the function defined in Lemma 1.2.7. Choose $\iota > 0$ so that $M - \varepsilon_- \ge \iota$, and set

$$f_{i,i}(x) = \sum_{k \in \mathbf{Z}^N} s_{i,k}(a_{i,k} * \Psi_i(x)),$$

$$f_{i,j}(x) = \sum_{k \in \mathbf{Z}^N} s_{i,k}(a_{i,k} * \psi_j(x)), \quad j > i.$$

Then, since $\mathcal{F}\Psi_i + \sum_{j=i+1}^\infty \mathcal{F}\psi_j = 1$,

$$f = \sum_{i=0}^\infty \sum_{j=i}^\infty f_{i,j} = \sum_{l=0}^\infty \sum_{j=l}^\infty f_{j-l,j}.$$

By Lemmas 1.2.9 and 1.2.7

$$|f_{j-l,j}(x)| \le \sum_{k \in \mathbf{Z}^N} |s_{j-l,k}| |(a_{j-l,k} * \psi_j(x))|$$

$$\le C 2^{-lM} \sum_{k \in \mathbf{Z}^N} |s_{j-l,k}| \gamma_{j-l}(x - k2^{-j+l})$$

$$\le C 2^{-lM} M_{r,d} g_{j-l}(x) \le C 2^{-l(\varepsilon_- + \iota)} M_{r,d} g_{j-l}(x),$$

and thus, setting $f_j = g_j = 0$ for $j < 0$, by Definition 1.1.1,

$$\left\|\{f_{j-l,j}\}_{j=0}^\infty\right\|_E \leq C 2^{-l(\varepsilon_- + \iota)} \left\|\{M_{r,d} g_{j-l}\}_{j=0}^\infty\right\|_E$$
$$= C 2^{-l(\varepsilon_- + \iota)} \left\|(S_-)^l \{M_{r,d} g_j\}_{j=0}^\infty\right\|_E \leq C 2^{-l\iota} \left\|\{g_j\}_{j=0}^\infty\right\|_E.$$

It follows from Definition 1.1.5 and the condition

$$\operatorname{supp} \mathcal{F} f_{j-l,j} \subset \operatorname{supp} \mathcal{F} \psi_j \subset B(0, 2^{j+1}),$$

that the function $F_l = \sum_{j=l}^\infty f_{j-l,j}$ belongs to the space $YL(E)$ and satisfies

$$\|F_l\|_{YL(E)} \leq \left\|\{f_{j-l,j}\}_{j=0}^\infty\right\|_E \leq C 2^{-j\iota} \left\|\{g_i\}_{i=0}^\infty\right\|_E.$$

It is a consequence of the quasi-norm property (1.1.2) and the completeness of $YL(E)$ that $f = \sum_{j=0}^\infty F_j$ belongs to $YL(E)$, and

$$\|f\|_{YL(E)} \leq C \left\|\{g_i\}_{i=0}^\infty\right\|_E.$$

Step 3. We now turn to the implication (ii*) $\Rightarrow$ (v). By Lemma 1.2.1 there are polynomials $\pi_{i,k} \in \mathfrak{P}_{M-1}$ such that

$$|f(x) - \pi_{i,k}(x)| \chi_{Q_{i,k}(3)}(x)$$
$$\leq C \left(2^{iN} \int_{|z| < 2^{-i}} |\Delta_z^M f(x)|^q \, dz \right)^{1/q}$$
$$+ C \left(2^{2iN} \int_{Q_{i,k}(6)} \int_{|z| < 2^{-i}} |\Delta_z^M f(y)|^q \, dz \, dy \right)^{1/q}.$$

If functions g_i are defined by (1.1.22) and (1.1.23), it follows that

$$|f(x) - \pi_{i,k}(x)| \chi_{Q_{i,k}(3)}(x) \leq C(g_i(x) + M_q g_i(x)) \leq C M_r g_i(x) \quad \text{for } i \geq 1,$$

and $|f(x)| \leq C M_r g_0(x)$. Since any x belongs to a bounded number of cubes $Q_{i,k}(3)$ for each i, the desired implication follows from the definition of $S(\varepsilon_+, \varepsilon_-, r)$.

Step 4. The claim (v) $\Rightarrow$ (iii*) follows if we define g_i by (1.1.28), let $x \in Q_{i,k}$, and observe that then

$$2^{iN/q} \mathcal{E}_M(f, B(x, 2^{-i}), L_q)) \leq 2^{iN/q} \mathcal{E}_M(f, Q_{i,k}(3), L_q)) \leq C M_q g_i(x)) \leq C M_r g_i(x)$$

for $i > 0$, and $\|f\|_{L_q(B(x,1))} \leq C M_r g_0(x)$.

Step 5. To prove that (iii*) $\Rightarrow$ (iv) we denote by $\pi_{i,k}$ polynomials in $\mathfrak{P}_{M-1}$ such that

$$\|f - \pi_{i,k}\|_{L_q(Q_{i,k}(3))} \leq 2 \mathcal{E}_M(f, Q_{i,k}(3), L_q), \quad (i, k) \in \mathbf{N}_0 \times \mathbf{Z}^N.$$

Let φ be a function satisfying (1.3.1), set $\varphi_{i,k}(x) = \varphi(2^i x - k)$, and let i_0 be some number in $\mathbf{N}_0$. Define functions $u_{i,k}(x)$ for $i \geq i_0$ by

(1.3.3) $$u_{i,k} = \varphi_{i,k} \sum_{l \in \mathbf{Z}^N} \varphi_{i-1,l}(\pi_{i,k} - \pi_{i-1,l}), \quad i \geq i_0 + 1,$$

(1.3.4) $$u_{i_0,k} = \varphi_{i_0,k} \pi_{i_0,k},$$

and set $u_{i,k} = 0$ for $i < i_0$. Then

$$\sum_{k \in \mathbf{Z}^N} u_{i,k} = \sum_{k \in \mathbf{Z}^N} \pi_{i,k} \varphi_{i,k} \sum_{l \in \mathbf{Z}^N} \varphi_{i-1,l} - \sum_{k \in \mathbf{Z}^N} \varphi_{i,k} \sum_{l \in \mathbf{Z}^N} \pi_{i-1,l} \varphi_{i-1,l}$$

$$= \sum_{k \in \mathbf{Z}^N} \pi_{i,k} \varphi_{i,k} - \sum_{l \in \mathbf{Z}^N} \pi_{i-1,l} \varphi_{i-1,l}, \quad i \geq i_0 + 1.$$

This identity and Lemma 1.2.3 imply that

$$f(x) = \sum_{i=0}^{\infty} \sum_{k \in \mathbf{Z}^N} u_{i,k}(x) = \lim_{i \to \infty} \sum_{k \in \mathbf{Z}^N} \pi_{i,k}(x) \varphi_{i,k}(x)$$

with convergence in $L_{r,\mathrm{loc}}$, and pointwise absolute convergence almost everywhere. Moreover, the function

$$\sum_{(i,k) \in \mathbf{N}_0 \times \mathbf{Z}^N} |u_{i,k}(x)|$$

belongs to $L_{r,\mathrm{loc}}$.

In order to estimate $D^\alpha u_{i,k}$ for $|\alpha| \leq M$ we observe that for $i > i_0$ and a sufficiently large i_0 by Lemma 1.2.2

$$\|D^\alpha(\pi_{i,k} - \pi_{i-1,l})\|_{L_\infty(Q_{i,k}(3))} \leq C 2^{i|\alpha|} 2^{iN/q} \|\pi_{i,k} - \pi_{i-1,l}\|_{L_q(Q_{i,k}(3))}$$
$$\leq C 2^{i|\alpha|} 2^{iN/q} \mathcal{E}_M(f, B(x, 2^{-(i-i_0)}), L_q)$$

for all $x \in \mathrm{supp}\, u_{i,k}$, and thus by the Leibniz rule

$$\|D^\alpha u_{i,k}\|_{L_\infty(Q_{i,k}(3))} \leq C 2^{i|\alpha|} 2^{iN/q} \mathcal{E}_M(f, B(x, 2^{-(i-i_0)}), L_q)$$

for all $x \in Q_{i,k}(3)$. For $i = i_0$ similarly

$$\|D^\alpha u_{i_0,k}\|_{L_\infty(Q_{i_0,k}(3))} \leq C \|f\|_{L_q(Q_{i_0,k}(3))} \leq C \|f\|_{L_q(B(x,1))}, \quad x \in Q_{i_0,k}(3).$$

If we define $s_{i,k}$ by

$$s_{i,k} = \max_{|\alpha| \leq M} 2^{-i|\alpha|} \|D^\alpha u_{i,k}\|_{L_\infty(Q_{i,k}(3))},$$

and set

$$a_{i,k} = \frac{u_{i,k}}{s_{i,k}} \quad \text{if } s_{i,k} \neq 0, \quad a_{i,k} = 0 \quad \text{if } s_{i,k} = 0,$$

it follows that f has a representation

$$f(x) = \sum_{i=0}^{\infty} \sum_{k \in \mathbf{Z}^N} s_{i,k} a_{i,k}(x)$$

with all the desired properties.

Step 6. In the final two steps we prove that (iv) implies (ii) and (iii) for any $p \leq \infty$ such that $\frac{N}{p} > \frac{N}{r} - \epsilon_+$. Let

$$f(x) = \sum_{i=0}^{\infty} h_i(x), \quad h_i(x) = \sum_{k \in \mathbf{Z}^N} s_{i,k} a_{i,k}(x)$$

be a representation of f that satisfies the conditions in (iv). Then, for the functions g_i, $i \in \mathbf{N}$, appearing in (ii) in the theorem, we have the estimate

$$g_i(x) = \left(2^{iN}\int_{B(x,2^{-i})}\left|\sum_{j=0}^{\infty}\Delta_z^M h_j(x)\right|^p dz\right)^{1/p}$$

$$\leq C\left(2^{iN}\int_{B(x,2^{-i})}\left|\sum_{j=0}^{i-1}\Delta_z^M h_j(x)\right|^p dz\right)^{1/p}$$

$$+ C\left(2^{iN}\int_{B(x,2^{-i})}\left|\sum_{j=i}^{\infty}\Delta_z^M h_j(x)\right|^p dz\right)^{1/p}.$$

For $j \geq i$ we use the fact that

$$\Delta_z^M h_j(x) = (-1)^M h_j(x) + \sum_{k=1}^{M}(-1)^{M-k}\binom{M}{k}h_j(x+kz).$$

This gives

$$\left(2^{iN}\int_{B(x,2^{-i})}\left|\sum_{j=i}^{\infty}\Delta_z^M h_j(x)\right|^p dz\right)^{1/p}$$

$$\leq C\sum_{j=i}^{\infty}|h_j(x)| + C\left(2^{iN}\int_{B(x,M2^{-i})}\left(\sum_{j=i}^{\infty}|h_j(z)|\right)^p dz\right)^{1/p}.$$

Using the estimate

$$|h_j(x)| \leq \sum_l |s_{j,l}|\tilde{\chi}_{j,l}(x),$$

where $\tilde{\chi}_{j,l} = \chi(Q_{j,l}(3))$, we can apply Lemma 1.2.4 to estimate this last quantity.

For $j < i$, on the other hand, we use the estimate

$$|\Delta_z^M h_j(x)| \leq C|z|^M \sup_{|\alpha|\leq M}\sup_{y\in B(x,M2^{-i})}|D^\alpha h_j(y)| \leq C2^{-iM}2^{jM}\sum_l |s_{j,l}|,$$

where the summation is taken over all l such that $Q_{j,l}(3)$ intersects $B(x,M2^{-i})$. The desired result now follows from Lemma 1.2.6.

Step 7. The implication (iv) $\Rightarrow$ (iii) is proved similarly. With g_i defined as in (iii) we have

$$g_i(x) = 2^{iN/p}\mathcal{E}_M\left(\sum_{j=0}^{\infty}h_j, B(x,2^{-i}), L_p\right)$$

$$\leq C2^{iN/p}\mathcal{E}_M\left(\sum_{j=0}^{i-1}h_j, B(x,2^{-i}), L_p\right) + C2^{iN/p}\mathcal{E}_M\left(\sum_{j=i}^{\infty}h_j, B(x,2^{-i}), L_p\right).$$

For $j \geq i$ we use the trivial inequality

$$2^{iN/p}\mathcal{E}_M\left(\sum_{j=i}^{\infty}h_j, B(x,2^{-i}), L_p\right) \leq \left(2^{iN}\int_{B(x,2^{-i})}\left(\sum_{j=i}^{\infty}|h_j(z)|\right)^p dz\right)^{1/p},$$

and Lemma 1.2.4 as before.

For $j < i$ we have by Taylor's formula

$$2^{iN/p}\mathcal{E}_M\left(\sum_{j=0}^{i-1}h_j, B(x,2^{-i}), L_p\right)$$

$$\leq C2^{-iM}\sup_{|\alpha|\leq M}\sup_{y\in B(x,2^{-i})}\left|\sum_{j=0}^{i-1}D^\alpha h_j(y)\right| \leq C2^{-iM}2^{jM}\sum_{j,l}|s_{j,l}|,$$

1.4. Some Lemmas on Orthogonalization

For the proof of Theorem 1.1.15 we again need some auxiliary results. The first lemma is required for the proof of the implication (i) ⇒ (iii).

As before $Q_{0,0}$ denotes the unit cube $[0,1)^N$, and $Q_{j,k} = 2^{-j}(k + Q_{0,0})$ for $k = (k_1, k_2, \ldots, k_N) \in \mathbf{Z}^N$.

LEMMA 1.4.1. *Let $M, L \in \mathbf{N}_0$, let $\kappa \in C^\infty$, $\kappa \perp \mathfrak{P}_L$, and suppose that for some $\lambda \geq L + N + 1$*

$$\sup_y (1 + |y|)^\lambda |D^\alpha \kappa(y)| \leq 1, \quad \textit{for all } \alpha \textit{ with } |\alpha| \leq M.$$

Then there are C^M functions $\{b_k\}_{k \in \mathbf{Z}^N}$, and constants S and C depending on M, L, and N, such that if $\lambda > S$, then

(i) $\sum_{k \in \mathbf{Z}^N} b_k = \kappa$,
(ii) $\operatorname{supp} b_k \subset Q_{0,k}(2)$,
(iii) $b_k \perp \mathfrak{P}_L$,
(iv) $\max_y |D^\alpha b_k(y)| \leq C(1 + |k|)^{S-\lambda}$ *for all α with $|\alpha| \leq M$.*

This is an easy consequence of the following three lemmas.

LEMMA 1.4.2. *Let M, L, κ, and λ be as above. Then there exist C^M functions κ_n for $n \in \mathbf{N}_0$, and a constant C depending on M, L, and N, such that*

(i) $\sum_{n=0}^\infty \kappa_n = \kappa$,
(ii) $\operatorname{supp} \kappa_0 \subset B(0,1)$, $\operatorname{supp} \kappa_n \subset B(0, 2^n) \setminus B(0, 2^{n-2})$ *for $n \geq 1$,*
(iii) $\kappa_n \perp \mathfrak{P}_L$,
(iv) $\max_y |D^\alpha \kappa_n(y)| \leq C 2^{-n\lambda}$ *for all α with $|\alpha| \leq M$.*

PROOF. Let $\Omega \in C_0^\infty(B(0,1))$ and assume that $\Omega(x) = 1$ on $B(0, \frac{1}{2})$. Define $\Omega_n(x) = \Omega(2^{-n}x)$, and set $\omega_n = \Omega_n - \Omega_{n-1}$ for $n \geq 1$, and $\omega_0 = \Omega$. Then $\sum_{n=0}^\infty \omega_n(x) \equiv 1$ and $\operatorname{supp} \omega_n \subset B(0, 2^n) \setminus B(0, 2^{n-2})$ for $n \geq 1$.

For multi-indices $\alpha \in \{0, 1, \ldots, L\}^N$, we let

$$\{e_\alpha : |\alpha| \leq L\}, \qquad e_\alpha \in C_0^\infty(B(0,1) \setminus B(0, \tfrac{1}{2}))$$

be a system dual to the system of monomials $\{x^\beta : |\beta| \leq L\}$, i.e.,

$$\int_{\mathbf{R}^N} e_\alpha x^\alpha \, dx = 1, \qquad \int_{\mathbf{R}^N} e_\alpha x^\beta \, dx = 0 \quad \text{for } \alpha \neq \beta.$$

We first take the existence of such a dual system for granted, and return to this question at the end of the proof.

For $n \in \mathbf{N}_0$ we set

$$e_{\alpha,n}(x) = 2^{-n(|\alpha|+N)} e_\alpha(x 2^{-n}),$$

so that

$$e_{\alpha,n} \in C_0^\infty(B(0, 2^n) \setminus B(0, 2^{n-2})),$$

and

$$\int_{\mathbf{R}^N} e_{\alpha,n} x^\alpha \, dx = 1, \qquad \int_{\mathbf{R}^N} e_{\alpha,n} x^\beta \, dx = 0 \quad \text{for } \alpha \neq \beta.$$

We now set

$$\kappa_0 = \kappa\omega_0 - \sum_{|\alpha|\leq L} e_{\alpha,0} \int_{\mathbf{R}^N} \kappa\omega_0 x^\alpha \, dx,$$

$$\kappa_1 = \kappa\omega_1 - \sum_{|\alpha|\leq L} e_{\alpha,1} \int_{\mathbf{R}^N} \kappa\omega_1 x^\alpha \, dx + \sum_{|\alpha|\leq L} (e_{\alpha,0} - e_{\alpha,1}) \int_{\mathbf{R}^N} \kappa\omega_0 x^\alpha \, dx,$$

$$\kappa_n = \kappa\omega_n - \sum_{|\alpha|\leq L} e_{\alpha,n} \int_{\mathbf{R}^N} \kappa\omega_n x^\alpha \, dx + \sum_{|\alpha|\leq L} (e_{\alpha,n-1} - e_{\alpha,n}) \sum_{i=0}^{n-1} \int_{\mathbf{R}^N} \kappa\omega_i x^\alpha \, dx.$$

It follows that

$$\sum_{n=0}^{k} \kappa_n = \kappa \sum_{n=0}^{k} \omega_n - \sum_{|\alpha|\leq L} e_{\alpha,k} \int_{\mathbf{R}^N} \kappa x^\alpha \sum_{n=0}^{k} \omega_n \, dx,$$

which clearly converges to κ, as k tends to infinity, since $\kappa \perp \mathfrak{P}_L$. It is also easily seen that $\operatorname{supp}\kappa_0 \subset B(0,1)$, $\operatorname{supp}\kappa_n \subset B(0,2^n) \setminus B(0,2^{n-2})$ for $n \geq 1$, and that $\kappa_n \perp \mathfrak{P}_L$. Finally, to show (iv) we observe the identity

$$\sum_{i=0}^{n-1} \int_{\mathbf{R}^N} \kappa\omega_i x^\alpha \, dx = -\int_{\mathbf{R}^N} \kappa x^\alpha \sum_{i=n}^{\infty} \omega_i \, dx,$$

which again follows from the fact that $\kappa \perp \mathfrak{P}_L$. Here the right hand side is majorized by

$$\int_{|x|>2^{n-2}} |\kappa x^\alpha| \, dx \leq C \int_{|x|>2^{n-2}} |x|^{|\alpha|-\lambda} \, dx \leq C 2^{n(|\alpha|-\lambda+N)},$$

with a C independent of λ, if $\lambda \geq |\alpha| + N + 1$, and (iv) follows.

It remains to prove the existence of the dual basis $\{e_\alpha\}$. It is enough to construct functions $e_\alpha \in C_0^\infty(Q)$, where $Q \subset B(0,1) \setminus B(0,\frac{1}{2}))$ is a cube. By a change of variables we can assume that Q is the unit cube.

Let $0 < a_0 < a_1 < \ldots < a_L < 1$, and set $a_\gamma = \{a_{\gamma_0}, a_{\gamma_1}, \ldots, a_{\gamma_N}\}$ for all $\gamma = \{\gamma_0, \gamma_1, \ldots, \gamma_N\} \in \{0,1,\ldots,L\}^N$, so that we obtain L^N points a_γ in Q. We first prove that there are linear combinations e_α of Dirac measures at these points,

$$e_\alpha = \sum_{\gamma \in \{0,1,\ldots,L\}^N} c_{\alpha,\gamma} \delta(a_\gamma), \quad \alpha \in \{0,1,\ldots,L\}^N,$$

satisfying $\langle e_\alpha, x^\alpha \rangle = 1$ and $\langle e_\alpha, x^\beta \rangle = 0$ for all $\alpha, \beta \in \{0,1,\ldots,L\}^N$ with $\alpha \neq \beta$. Since $\langle \delta(a_\gamma), x^\beta \rangle = (a_\gamma)^\beta$ it is enough to prove that the linear system in L^N variables,

$$\sum_{\gamma \in \{0,1,\ldots,L\}^N} (a_\gamma)^\beta y_\gamma = 0, \quad \beta \in \{0,1,\ldots,L\}^N,$$

has only the trivial solution $y_\gamma = 0$, $\gamma \in \{0,1,\ldots,L\}^N$.

Equivalently, the transposed system

$$(1.4.1) \qquad \sum_{\beta \in \{0,1,\ldots,L\}^N} (a_\gamma)^\beta z_\beta = 0, \quad \gamma \in \{0,1,\ldots,L\}^N,$$

has only the trivial solution. But this is easily proved, for example by induction on N. In fact, for $N = 1$ the system (1.4.1) reads

$$\sum_{\beta=0}^{L} a_\gamma^\beta z_\beta = 0, \quad \gamma \in \{0, 1, \ldots, L\},$$

and the existence of a non-trivial solution means that the polynomial $p(t) = \sum_{\beta=0}^{L} z_\beta t^\beta$ has $L + 1$ different roots. (Alternatively, one observes that the matrix of the system is of Vandermonde type.) Assume now that the claim is true for $N - 1$ variables. For $\gamma = \{\gamma_0, \gamma_1, \ldots, \gamma_N\} \in \{0, 1, \ldots, L\}^N$ we denote $\gamma = \{\gamma', \gamma_N\}$, where $\gamma' = \{\gamma_0, \gamma_1, \ldots, \gamma_{N-1}\} \in \{0, 1, \ldots, L\}^{N-1}$, and similarly for β. Then

$$\sum_{\beta \in \{0,1,\ldots,L\}^N} (a_\gamma)^\beta z_\beta = \sum_{\beta_N = 0}^{L} \left(\sum_{\beta' \in \{0,1,\ldots,L\}^{N-1}} (a_{\gamma'})^{\beta'} z_{\beta', \beta_N} \right) a_{\gamma_N}^{\beta_N},$$

so that if there is a non-trivial solution to (1.4.1), it follows from the result for $N = 1$ that

$$\sum_{\beta' \in \{0,1,\ldots,L\}^{N-1}} (a_{\gamma'})^{\beta'} z_{\beta', \beta_N} = 0$$

for all $\beta_N \in \{0, 1, \ldots, L\}$ and all $\gamma' \in \{0, 1, \ldots, L\}^{N-1}$, which is impossible by the induction hypothesis.

Finally, we can "fatten" the Dirac measures $\delta(a_\gamma)$ by convolution with a C_0^∞ approximate identity. If the support is sufficiently small, the matrix of the modified coefficients $(a_\gamma)^\beta$ in (1.4.1) will still be invertible, and the modified e_α will belong to $C_0^\infty(Q)$ as claimed. $\square$

LEMMA 1.4.3. *Let $\eta \in C_0^M(Q_{0,0})$, $\eta \perp \mathfrak{P}_L$, and $\max_y |D^\alpha \eta(y)| \leq 1$ for all α with $|\alpha| \leq M$. Then there exist C^M functions η_i for $i \in \{0, 1\}^N$, and a constant A depending on M, L, and N, such that*

(i) $\sum_{i \in \{0,1\}^N} \eta_i = \eta$,
(ii) $\operatorname{supp} \eta_i \in Q_{1,i}(\frac{3}{2})$,
(iii) $\eta_i \perp \mathfrak{P}_L$,
(iv) $\max_y |D^\alpha \eta_i(y)| \leq A$ *for all α with $|\alpha| \leq M$.*

PROOF. For all multi-indices α with $|\alpha| \leq L$ we choose functions $u_\alpha \in C_0^M$ such that $\operatorname{supp} u_\alpha \subset [\frac{3}{8}, \frac{5}{8}]^N$, $\int_{\mathbf{R}^N} u_\alpha x^\beta \, dx = 0$ for all multi-indices $\beta \neq \alpha$, $|\beta| \leq L$, and $\int_{\mathbf{R}^N} u_\alpha x^\alpha \, dx = 1$.

Let φ_i, $i \in \{0, 1\}^N$, be C^M functions such that $\operatorname{supp} \varphi_i \subset Q_{1,i}(\frac{3}{2})$, and $\sum_{i \in \{0,1\}^N} \varphi_i(x) = 1$ on $Q_{0,0}$.

If η_i is defined by

$$\eta_i = \eta \varphi_i - \sum_{|\alpha| \leq L} u_\alpha \int_{\mathbf{R}^N} \eta \varphi_i x^\alpha \, dx,$$

it is easily seen that η_i has all the desired properties. $\square$

LEMMA 1.4.4. *Let η satisfy the assumptions of Lemma 1.4.3. Then, for every $j \in \mathbf{N}$ there exist C^M functions σ_k for $k \in \{0, 1, \ldots, 2^j - 1\}^N$, and constants C and S, depending on M, L, and N, such that*

(i) $\sum_{k \in \{0,1,\ldots,2^j-1\}^N} \sigma_k = \eta$,
(ii) $\operatorname{supp} \sigma_k \in Q_{j,k}(2)$,

(iii) $\sigma_k \perp \mathfrak{P}_L$,
(iv) $\max_y |D^\alpha \sigma_k(y)| \leq C 2^{Sj}$ for all α with $|\alpha| \leq M$.

PROOF. Let the functions η_i, $i \in \{0,1\}^N$, and the constant A be as in Lemma 1.4.3. Now apply the same lemma again to each of the 2^N functions η_i to obtain 2^{2N} functions η_{i_1,i_2} with support contained in cubes with side $(\frac{3}{4})^2$, and $\max_y |D^\alpha \eta_{i_1,i_2}(y)| \leq A^2 (\frac{4}{3})^{|\alpha|}$. After repeating this procedure s times we obtain 2^{sN} functions η_i, $i = (i_1, i_2, \ldots, i_s)$, with support contained in cubes with side $(\frac{3}{4})^s$, satisfying $\sum \eta_i = \eta$, $\eta_i \perp \mathfrak{P}_L$, and
$$\max_y |D^\alpha \eta_i(y)| \leq A^s (\tfrac{4}{3})^{s|\alpha|} \quad \text{for all } |\alpha| \leq M.$$
It is easily checked that each η_i has its support contained in $Q_{0,0}(2)$.

Now let $j \in \mathbf{N}$ be given, and choose s so that $2^{-j-2} < (\frac{3}{4})^s \leq 2^{-j-1}$. Then partition the set of indices i into disjoint subsets Γ_k for such $k \in \mathbf{Z}^N$ that $\operatorname{supp} \eta_i \subset Q_{j,k}(2)$ if $i \in \Gamma_k$. This is possible by the choice of s. Then set $\sigma_k = \sum_{i \in \Gamma_k} \eta_i$, so that $\operatorname{supp} \sigma_k \subset Q_{j,k}(2) \subset Q_{0,0}(2)$, and $\sum_k \sigma_k = \eta$. The number of elements in Γ_k is at most 2^{sN}, and it follows from the choice of s that there are constants C and S such that
$$\max_y |D^\alpha \sigma_k(y)| \leq 2^{sN} (\tfrac{4}{3})^{s|\alpha|} A^s \leq C 2^{Sj} \quad \text{for all } |\alpha| \leq M.$$
The lemma follows. □

PROOF OF LEMMA 1.4.1. The functions κ_n in Lemma 1.4.2 have their supports contained in squares of side 2^{n+1}. After a change of scale and a translation we can apply Lemma 1.4.4 to each of the κ_n. We obtain functions $\sigma_{n,k} \in C_0^M(Q_{0,k}(2))$, such that
$$\max_y |D^\alpha \sigma_{n,k}(y)| \leq C 2^{-n\lambda} 2^{Sn}, \quad |\alpha| \leq M,$$
and $\kappa_n = \sum_{k \in \mathbf{Z}^N} \sigma_{n,k}$, where we observe that $\sigma_{n,k} = 0$ for those k and n for which $Q_{0,k}$ is not contained in $B(0, 2^{n+1})$. Setting $b_k = \sum_{n=0}^\infty \sigma_{n,k}$, the estimate (iv) in the lemma follows easily for any $\lambda > S$, and (i), (ii) and (iii) are obvious. □

The following lemma is needed for the proof of the implication (ii′) ⇒ (iii) in Theorem 1.1.15.

LEMMA 1.4.5. *Let $L \in \mathbf{N}$. Then there are functions η, κ, and ω, such that $\eta \in C_0^\infty(B(0, \frac{1}{2}))$, $\int_{\mathbf{R}^N} \eta(x)\, dx = 1$,*

(1.4.2) $$\kappa, \omega \in C_0^\infty(B(0,1)), \quad \kappa, \omega \perp \mathfrak{P}_L,$$

and

(1.4.3) $$\eta - 2^{-N} \eta(\cdot/2) = \kappa * \omega.$$

PROOF. Cf. Frazier and Jawerth [19], Theorem 2.6. Let u be a radial function (i.e., $u(x) = u(y)$ if $|x| = |y|$) belonging to $C_0^\infty(B(0, \frac{1}{4}))$, such that

(1.4.4) $$\int_{\mathbf{R}^N} u(x)\, dx = 1, \quad \text{and} \quad \int_{\mathbf{R}^N} u(x) x^\alpha\, dx = 0$$

for all multi-indices $\alpha \in \mathbf{N}_0^N$ with $0 < |\alpha| \leq 4L$. (We shall show shortly how to construct such a function.) Set $\eta = u * u$. Let E be the fundamental solution of the

Laplace equation, i.e., $E(x) = c_2 \log|x|$ if $N = 2$, and $E(x) = -c_N|x|^{2-N}$ if $N \geq 3$, so that $E * f = \Delta^{-1} f$ for $f \in \mathcal{S}$. We define the functions κ and ω by

$$\kappa = \Delta^L(u + 2^{-N} u(\cdot/2)),$$

and

$$\omega = \Delta^{-L}(u - 2^{-N} u(\cdot/2)).$$

The equality (1.4.3) follows immediately from the elementary identity

$$\eta - 2^{-N}\eta(\cdot/2) = (u + 2^{-N}u(\cdot/2)) * (u - 2^{-N}u(\cdot/2)).$$

We have to prove that the functions so constructed satisfy all the conditions imposed in the lemma. It is clear that $\eta \in C_0^\infty(B(0, \frac{1}{2}))$, and also that $\kappa \in C_0^\infty(B(0, \frac{1}{2})))$, and $\kappa \perp \mathfrak{P}_L$. To prove that $\omega \in C_0^\infty(B(0, \frac{1}{2}))$, we set

$$\omega_j(x) = \Delta^{-j}(u(x) - 2^{-N}u(x/2)) \quad \text{for } j = 0, 1, \ldots, L.$$

Denoting the Fourier transform $\mathcal{F}\omega_j$ by $\hat{\omega}_j$ we observe that

(1.4.5) $$\hat{\omega}_j(\xi) = |\xi|^{-2j}(\hat{u}(\xi) - \hat{u}(2\xi)) \quad \text{for } \xi \neq 0.$$

It follows from (1.4.4) that $\hat{u}(\xi) = 1 + O(|\xi|^{4L+1})$, as $\xi \to 0$, so we can set $\hat{\omega}_j(0) = \lim_{\xi \to 0} \hat{\omega}_j(\xi) = 0$, whence $\int_{\mathbf{R}^N} \omega_j(x) \, dx = 0$. Moreover, ω_j is radial.

For an $x \notin B(0, \frac{1}{2})$ we have

$$\omega_1(x) = E * \omega_0(x) = \int_{\mathbf{R}^N} E(x-y)\omega_0(y) \, dy = E(x) \int_{\mathbf{R}^N} \omega_0(y) \, dy = 0,$$

by the spherical symmetry of ω_0, and the mean value property of harmonic functions. Similarly, $\omega_{j+1} = E * \omega_j$, and it follows by iteration that $\omega = \omega_L \in C_0^\infty(B(0, \frac{1}{2}))$.

It also follows from (1.4.4) and (1.4.5) that $D^\alpha \hat{\omega}(0) = 0$ for $|\alpha| \leq L$, and thus $\int_{\mathbf{R}^N} \omega(x) x^\alpha \, dx = 0$, i.e., $\omega \perp \mathfrak{P}_L$.

In order to finish the proof it remains to construct a function u so that (1.4.4) is satisfied. Let v_0 be a radial function belonging to $C_0^\infty(B(0, 2^{-4}/L))$ such that $\int_{\mathbf{R}^N} v_0(x) \, dx = 1$. Then we define u by setting $u = v_J$, where

$$v_{j+1} = 2v_j - v_j * v_j \quad \text{for } j = 0, 1, \ldots, J,$$

and J is chosen so that $2^J \leq 4L < 2^{J+1}$. Clearly u is radial and belongs to $C_0^\infty(B(0, \frac{1}{4}))$. Also $\int_{\mathbf{R}^N} u(x) \, dx = \hat{u}(0) = 1$, since $\hat{v}_0(0) = 1$, and $\hat{v}_{j+1} = 2\hat{v}_j - \hat{v}_j^2$ for each j.

If we denote $\hat{v}_j = 1 - \delta_j$, the last equation becomes $\delta_{j+1} = \delta_j^2$, whence $\delta_j = \delta_0^{2^j}$ for all j. The smoothness and spherical symmetry of $\hat{v}_0$ imply that $\delta_0(\xi) = O(|\xi|^2)$, and thus $\delta_j(\xi) = O(|\xi|^{2^{j+1}})$, as $\xi \to 0$. If $2^{J+1} > 4L$ it follows that $u = v_J$ satisfies $D^\alpha \hat{u}(0) = 0$, and hence also $\int_{\mathbf{R}^N} u(x) x^\alpha \, dx = 0$ for all multi-indices α with $0 < |\alpha| \leq 4L$, as claimed. $\square$

1.5. Proof of Theorem 1.1.15

We again organize the proof in steps. We prove the implications (ii) $\Rightarrow$ (i), (iii) $\Rightarrow$ (ii), (ii') $\Rightarrow$ (iii), and (i) $\Rightarrow$ (iii). The implication (ii) $\Rightarrow$ (ii') is trivial.

Step 1. (ii) $\Rightarrow$ (i). Let $f \in \mathcal{S}'$, let Ψ be defined as in (1.3.2), set $\Psi_{2^{-j}} = 2^{jN}\Psi(\cdot 2^j)$ for $j \in \mathbf{N}_0$, $\psi_0 = \Psi$, and $\psi_j = \Psi_{2^{-j}} - \Psi_{2^{-(j-1)}}$ for $j \in \mathbf{N}$, so that $\sum_{j=0}^n \psi_j = \Psi_{2^{-n}}$, and $\lim_{n\to\infty} \mathcal{F}\Psi_{2^{-n}}(\xi) \equiv 1$. It is easily seen that $f * \Psi_{2^{-n}} \to f$ in the sense of $\mathcal{S}'$, as $n \to \infty$. Thus, setting $f_j = f * \psi_j$, we have $f = \sum_{j=0}^{\infty} f_j$, with convergence in $\mathcal{S}'$. Here $\mathrm{supp}\,\mathcal{F}f_j \subset \mathrm{supp}\,\mathcal{F}\psi_j \subset B(0, 2^{j+1}) \setminus B(0, 2^{j-1})$ for $j \geq 1$, and $\mathrm{supp}\,\mathcal{F}f_0 \subset B(0, 2)$.

Now assume that f satisfies the conditions in (ii). For any $\lambda \geq 0$ and any positive integer L there is a constant C such that
$$\max_x (1+|x|)^\lambda |D^\alpha \Psi(x)| \leq C \quad \text{for all } \alpha \text{ with } |\alpha| \leq L.$$
Since all moments of ψ_j vanish for $j \geq 1$, it follows from (1.1.30) that for the same constant C, the functions $f_j(x) = f * \psi_j(x) = \langle f, \psi_j(x-\cdot)\rangle$ satisfy $|f_j| \leq Ch_j$, with h_j defined as in (ii). The claim follows.

Step 2. (iii) $\Rightarrow$ (ii). We prove the implication for $0 < r < 1$. For $r \geq 1$ the proof simplifies, and is omitted. Assume that f has a representation $f = \sum_{i,k} s_{i,k} a_{i,k}$, satisfying the conditions in (iii); in particular the functions $g_i = \sum_{k \in \mathbf{Z}^N} |s_{i,k}|\chi_{i,k}$ satisfy $\|\{g_i\}_{i=0}^\infty\|_E < \infty$.

Let $\{h_i\}_{i=0}^\infty$ be the sequence of functions defined in (ii), let φ and ψ satisfy the conditions in (ii), in particular (1.1.30), and set $\psi_0(x) = \varphi(x)$, $\psi_j(x) = 2^{jN}\psi(2^j x)$ for $j \geq 1$. Then, for $j \geq 0$,
$$h_j(x) = \sup_{\psi_j} |\langle f, \psi_j(x-\cdot)\rangle| = \sup_{\psi_j} |f * \psi_j(x)|,$$
where the supremum is taken over all ψ or φ, satisfying (ii). On the other hand,
$$|f * \psi_j(x)| \leq \sum_{i=0}^\infty \sum_{k \in \mathbf{Z}^N} |s_{i,k}||a_{i,k} * \psi_j(x)|.$$
If we denote
$$\sum_{k \in \mathbf{Z}^N} |s_{i,k}| \sup_{\psi_j} |a_{i,k} * \psi_j(x)| = H_{i,j}(x), \quad i \geq 0, \; j \geq 0,$$
it follows that
$$h_j(x) \leq \sum_{i=0}^\infty H_{i,j}(x).$$
By (1.2.9) and (1.2.8) in Lemma 1.2.9, and with γ_j defined as in Lemma 1.2.7,
$$|a_{i,k} * \psi_j(x)| \leq \frac{C 2^{-(i-j)(L+N)}}{(1+2^j|x-k2^{-i}|)^\lambda} \leq C 2^{-(i-j)(L+N)} \gamma_j(x-k2^{-i}),$$
for $0 \leq j \leq i$, and
$$|a_{i,k} * \psi_j(x)| \leq \frac{C 2^{-(j-i)M}}{(1+2^i|x-k2^{-i}|)^\lambda} \leq C 2^{-(j-i)M} \gamma_i(x-k2^{-i}),$$
for $0 \leq i \leq j$. By Lemma 1.2.7
$$H_{i,j}(x) \leq C 2^{-(i-j)(L+N)} T_j g_i(x) \leq C 2^{-(i-j)(L+N-N/r)} M_{r,d} g_i(x)$$
for $0 \leq j \leq i$, and
$$H_{i,j}(x) \leq C 2^{-(j-i)M} T_i g_i(x) \leq C 2^{-(j-i)M} M_{r,d} g_i(x)$$
for $0 \leq i \leq j$.

Choose $\iota > 0$ so that $M - \varepsilon_- \geq \iota$ and $L + N - N/r + \varepsilon_+ \geq \iota$. It follows that
$$H_{i,j}(x) \leq C 2^{-(j-i)(\varepsilon_-+\iota)} M_{r,d} g_i(x), \quad 0 \leq i \leq j,$$
$$H_{i,j}(x) \leq C 2^{(i-j)(\varepsilon_+-\iota)} M_{r,d} g_i(x), \quad i \geq j.$$
Using the fact that
$$\left\{\sum_{i=0}^{\infty} H_{i,j}\right\}_{j=0}^{\infty} = \sum_{l=0}^{\infty} \{H_{j-l,j}\}_{j=l}^{\infty} + \sum_{l=1}^{\infty} \{H_{j+l,j}\}_{j=0}^{\infty},$$
and setting $g_i = 0$ for $i < 0$, we see that $\|\{h_j\}_{j=0}^{\infty}\|_E$ does not exceed the norm of
$$\sum_{l=0}^{\infty} 2^{-l(\varepsilon_-+\iota)} \{M_{r,d} g_{j-l}(x)\}_{j=0}^{\infty} + \sum_{l=1}^{\infty} 2^{l(\varepsilon_+-\iota)} \{M_{r,d} g_{j+l}(x)\}_{j=0}^{\infty}.$$
But
$$\|\{M_{r,d} g_{j-l}\}_{j=0}^{\infty}\|_E = \|S_-^l(\{M_{r,d} g_j\}_{j=0}^{\infty})\|_E \leq C 2^{l\varepsilon_-} \|\{g_j\}_{j=0}^{\infty}\|_E,$$
and
$$\|\{M_{r,d} g_{j+l}\}_{j=0}^{\infty}\|_E = \|S_+^l(\{M_{r,d} g_j\}_{j=0}^{\infty})\|_E \leq C 2^{-l\varepsilon_+} \|\{g_j\}_{j=0}^{\infty}\|_E,$$
whence
$$\|\{h_j\}_{j=0}^{\infty}\|_E^{\kappa} \leq C \|\{g_j\}_{j=0}^{\infty}\|_E^{\kappa} \sum_{l=0}^{\infty} 2^{-l\iota\kappa} = C \|\{g_j\}_{j=0}^{\infty}\|_E^{\kappa}.$$
The claim follows.

Step 3. (ii') $\Rightarrow$ (iii). Let $f \in \mathcal{S}'$ and suppose that the functions h_i defined in (ii') satisfy $\|\{h_i\}_{i=0}^{\infty}\|_E < \infty$. Let $\eta \in C_0^{\infty}(B(0, \frac{1}{2}))$ satisfy $\int_{\mathbf{R}^N} \eta \, dx = 1$, and let $\eta_i = 2^{iN} \eta(\cdot 2^i)$, so that $\operatorname{supp} \eta_i \subset B(0, 2^{-i-1})$. We can suppose that $\eta - \eta_{-1} = \kappa * \omega$ as in Lemma 1.4.5, where κ and ω have vanishing moments of order $< L$.

We know that $f * \eta_i \to f$ in $\mathcal{S}'$, as $i \to \infty$. We have
$$f * \eta_i = f * \eta_0 + \sum_{j=1}^{i} f * (\eta_j - \eta_{j-1}) = f * \eta_0 + \sum_{j=1}^{i} f * \kappa_j * \omega_j,$$
where κ_j and ω_j are defined in analogy with η_j, and have support in $B(0, 2^{-i-1})$. Then we can write
$$f * \kappa_j * \omega_j = \left(\sum_{k \in \mathbf{Z}^N} \chi_{j,k}(f * \kappa_j)\right) * \omega_j = \sum_{k \in \mathbf{Z}^N} ((f * \kappa_j) \chi_{j,k}) * \omega_j.$$
We define constants $s_{j,k}$ by setting $((f * \kappa_j) \chi_{j,k}) * \omega_j = s_{j,k} a_{j,k}$ and requiring the $a_{j,k}$ to be smooth atoms satisfying (1.1.26). This is possible, since clearly $\operatorname{supp}(\chi_{j,k} * \omega_j) \subset Q_{j,k}(3)$, and since $a_{j,k}$ has as many vanishing moments as ω_j. It follows that f has the desired representation $f = \sum s_{j,k} a_{j,k}$ with convergence in $\mathcal{S}'$, and it only remains to prove that the coefficients $s_{j,k}$ satisfy the required estimate.

For all x we have by assumption $|(f * \kappa_j)(x)| = |\langle f, \kappa_j(x - \cdot)\rangle| \leq h_j(x)$. Moreover, $\operatorname{supp} \kappa_j(x - \cdot) \subset B(y, 2^{-j+1})$ for any $y \in B(x, 2^{-j})$, and thus for $j \geq 1$ $|(f * \kappa_j)(x)| \leq 2^N h_{j-1}(y)$, if $y \in B(x, 2^{-j})$. It follows that for $x \in Q_{j,k}$
$$|(f * \kappa_j)(x)| \leq C 2^{jN/r} \left(\int_{B(x, 2^{-j})} h_{j-1}(y)^r \, dy\right)^{1/r}$$
$$\leq C 2^{jN/r} \left(\int_{B(z, 2^{-j}(1+\sqrt{N}))} h_{j-1}(y)^r \, dy\right)^{1/r}$$
for any $z \in Q_{j,k}$, and thus for a suitable constant C
$$\max_{x \in Q_{j,k}} |(f * \kappa_j)(x)| \leq C \inf_{x \in Q_{j,k}} M_{r,d} h_{j-1}(x).$$

It follows easily that $|s_{j,k}| \leq C \inf_{x \in Q_{j,k}} M_{r,d} h_{j-1}(x)$, and consequently,

$$\sum_{k \in \mathbf{Z}^N} |s_{j,k}| \chi_{j,k} \leq C M_{r,d} h_{j-1}, \quad j \geq 1.$$

For $j = 0$ one similarly obtains, observing that $\operatorname{supp} \eta \subset B(0, \frac{1}{2})$, that

$$\max_{x \in Q_{j,k}} |(f * \eta_0)(x)| \leq C \inf_{x \in Q_{j,k}} M_{r,d} h_0(x),$$

and thus

$$\sum_{k \in \mathbf{Z}^N} |s_{0,k}| \chi_{0,k} \leq C M_{r,d} h_0.$$

This proves (iii).

Step 4. (i) $\Rightarrow$ (iii). Suppose that $f \in Y(E)$, and let M and L be nonnegative integers such that $M > \varepsilon_-$, and $L > N \max\{\frac{1}{r} - 1, 0\} - \varepsilon_+$. Then f has a representation $f = \sum_{i=0}^{\infty} f_i$, satisfying (1.1.10) and (1.1.11), and converging in $\mathcal{S}'$, such that $\|\{f_i\}_{i=0}^{\infty}\|_E \leq 2 \|f\|_{Y(E)}$.

There is a $\psi \in \mathcal{S}$ such that $\operatorname{supp} \mathcal{F}\psi \subset B(0, 2(1+\delta)) \setminus B(0, 2^{-1}(1-\delta)))$ for some δ, $0 < \delta < 1$, and $\mathcal{F}\psi(\xi) = 1$ on $B(0,2) \setminus B(0, 2^{-1})$. Set $\psi_i(x) = 2^{iN} \psi(2^i x)$ for $i \geq 1$, so that $\mathcal{F}\psi_i(\xi) = 1$ on $B(0, 2^{i+1}) \setminus B(0, 2^{i-1})$.

Let $\psi_0 \in \mathcal{S}$ be such that $\operatorname{supp} \mathcal{F}\psi_0 \subset B(0, 2(1+\delta))$ and $\mathcal{F}\psi_0(\xi) = 1$ on $B(0,2)$. Then $f_i = f_i * \psi_i$ for $i \geq 0$.

Let $\omega \in C_0^{\infty}(Q(2))$ be nonnegative and such that $\sum_{k \in \mathbf{Z}^N} \omega(x - k) \equiv 1$, and set $\omega_{i,k}(x) = \omega(2^i x - k)$. Then $f_i = \sum_{k \in \mathbf{Z}^N} (f_i \omega_{i,k}) * \psi_i$ for $i \geq 0$.

For any $\lambda > 0$ there is C such that for all $|\alpha| \leq M$ and all x we have $|D^\alpha \psi(x)| \leq C(1 + |x|)^{-\lambda}$. Therefore we can write $(f_i \omega_{i,k}) * \psi_i(x) = m_{i,k} \kappa_{i,k}(x)$, where by Lemma 1.1.7

$$m_{i,k} \leq \max_{y \in Q_{i,k}(2)} |f_i(y)| \leq C \inf_{y \in Q_{i,k}} M_{r,d} f_i(y),$$

and $\kappa_{i,k}$ is a function in $\mathcal{S}$ with vanishing moments of all orders, which satisfies

$$|D^\alpha \kappa_{i,k}(x)| \leq C 2^{i|\alpha|} (1 + |2^i x - k|)^{-\lambda},$$

and consequently for $x \in Q_{i,l}(2)$,

$$|D^\alpha \kappa_{i,k}(x)| \leq C 2^{i|\alpha|} (1 + |k - l|)^{-\lambda}.$$

Moreover, for any $x \in Q_{i,l}(2)$, setting $\rho = 2^{-i}(2\sqrt{N} + |k - l|)$,

$$\inf_{y \in Q_{i,k}} M_{r,d} f_i(y) \leq \left(2^{iN} \int_{Q_{i,k}} M_{r,d} f_i(y)^r \, dy \right)^{1/r}$$

$$\leq C \left(\frac{(1 + |k - l|)^N}{\rho^N} \int_{B(x,\rho)} M_{r,d} f_i(y)^r \, dy \right)^{1/r}$$

$$\leq C (1 + |k - l|)^{N/r} (1 + 2^{-i} |k - l|)^d \left(\frac{1}{\rho^N} \int_{B(0, \rho)} \frac{M_{r,d} f_i(x+z)^r}{(1 + |z|)^{rd}} \, dz \right)^{1/r}$$

$$\leq C (1 + |k - l|)^{N/r + d} M_{r,d}(M_{r,d} f_i(x)).$$

It follows from Lemma 1.4.1 by means of a change of coordinates that for each $i \geq 1$ and $k \in \mathbf{Z}^N$ there exists a family of C^∞ functions $\{b_{i,k,l}\}_{l \in \mathbf{Z}^N}$ and constants

s and C depending on λ, M, L, and N, such that if $\lambda > s$, then

$$\sum_{l \in \mathbf{Z}^N} b_{i,k,l} = \kappa_{i,k},$$

$$\operatorname{supp} b_{i,k,l} \subset Q_{i,l}(2),$$

$$b_{i,k,l} \perp \mathfrak{P}_{L-1},$$

$$\max_{x \in Q_{i,l}(2)} |D^\alpha b_{i,k,l}(x)| \leq C 2^{i|\alpha|} (1 + |k - l|)^{s-\lambda}, \quad |\alpha| \leq M.$$

Set

$$a_{i,l}(x) = \sum_{k \in \mathbf{Z}^N} m_{i,k} b_{i,k,l}(x).$$

Then,

$$f_i(x) = \sum_{k \in \mathbf{Z}^N} m_{i,k} \kappa_{i,k}(x) = \sum_{k \in \mathbf{Z}^N} m_{i,k} \sum_{l \in \mathbf{Z}^N} b_{i,k,l}(x) = \sum_{l \in \mathbf{Z}^N} a_{i,l}(x),$$

$\operatorname{supp} a_{i,l} \subset Q_{i,l}(2)$, $a_{i,l} \perp \mathfrak{P}_{L-1}$, and for $x \in Q_{i,l}(2)$

$$|D^\alpha a_{i,k}(x)| \leq C 2^{i|\alpha|} \inf_{y \in Q_{i,l}} M_{r,d}(M_{r,d} f_i)(y) \sum_{k \in \mathbf{Z}^N} (1 + |k|)^{N/r + d + s - \lambda}.$$

Thus, if λ is chosen large enough, we have

$$\max_x |D^\alpha a_{i,k}(x)| \leq 2^{i|\alpha|} s_{i,k},$$

where the functions $g_i(x) = \sum_{k \in \mathbf{Z}^N} s_{i,k} \kappa_{i,k}(x)$ satisfy $\|\{g_i\}_{i=1}^\infty\|_E \leq C \|f\|_{Y(E)}$. The implication (i) $\Rightarrow$ (iii) follows. $\square$

1.6. Homogeneous Spaces

We will also study the "homogeneous" analogues of the spaces $Y(E)$ and $YL(E)$. In what follows we denote by $\dot{E}$ a quasi-Banach lattice of sequences $\{f_i\}_{i=-\infty}^\infty$ of Lebesgue measurable functions on $\mathbf{R}^N$. As before there is a number κ, $0 < \kappa \leq 1$, such that (1.1.2) holds true.

DEFINITION 1.6.1. Let $\varepsilon_+, \varepsilon_- \in \mathbf{R}$, $r > 0$. We will say that $\dot{E} \in \dot{S}(\varepsilon_+, \varepsilon_-, r)$ if the following conditions are satisfied:

(a) The shift operators S_+ and S_-, defined by

$$S_+(\{f_i\}_{i=-\infty}^\infty) = \{f_{i+1}\}_{i=-\infty}^\infty,$$
$$S_-(\{f_i\}_{i=-\infty}^\infty) = \{f_{i-1}\}_{i=-\infty}^\infty,$$

are continuous on $\dot{E}$, and they satisfy for all $j \in \mathbf{N}$ the inequalities

$$\|(S_+)^j(\{f_i\}_{i=-\infty}^\infty)\|_{\dot{E}} \leq C_1 2^{-j\varepsilon_+} \|\{f_i\}_{i=-\infty}^\infty\|_{\dot{E}},$$
$$\|(S_-)^j(\{f_i\}_{i=-\infty}^\infty)\|_{\dot{E}} \leq C_2 2^{j\varepsilon_-} \|\{f_i\}_{i=-\infty}^\infty\|_{\dot{E}}.$$

(b) The maximal operator $\mathcal{M}_r$, defined by

$$\mathcal{M}_r(\{f_i\}_{i=-\infty}^\infty) = \{M_r f_i\}_{i=-\infty}^\infty,$$

is bounded on $\dot{E}$.

The homogeneity has some advantages, but the double infinities also lead to certain difficulties. In the case of Besov spaces these were analyzed by Peetre in [**42**], pp. 50–57.

The following lemma is analogous to Lemma 1.1.4. Again, $\delta_{i,j}$ denotes the Kronecker symbol, so that $\{\delta_{0,j}\}_{i=-\infty}^{\infty} = \{\ldots, 0, \ldots, 0, 1, 0, \ldots, 0, \ldots\}$.

LEMMA 1.6.2. *Let $\dot{E} \in \dot{S}(\varepsilon_+, \varepsilon_-, r)$, and let $F = \{f_i\}_{i=-\infty}^{\infty} \in \dot{E}$. Suppose that $f_i \equiv 0$ for all $i > 0$. Then for any $M > \varepsilon_-$ there is C so that*

$$\big\|\{\delta_{0,j}\textstyle\sum_{i=-\infty}^{0} 2^{iM}|f_i|\}_{j=-\infty}^{0}\big\|_{\dot{E}} \leq C\|F\|_{\dot{E}}.$$

PROOF. We can assume that $f_i \geq 0$. Then

$$\{\delta_{0,j}\textstyle\sum_{i=-\infty}^{0} 2^{iM} f_i\}_{j=-\infty}^{\infty} = \sum_{i=-\infty}^{0} \{\delta_{0,j} 2^{iM} f_i\}_{j=-\infty}^{\infty}$$
$$= \sum_{i=-\infty}^{0} 2^{iM}(S_-)^{-i}\{\delta_{i,j} f_i\}_{j=-\infty}^{\infty}.$$

But by Definition 1.6.1(a)

$$\big\|(S_-)^{-i}\{\delta_{i,j} f_i\}_{j=-\infty}^{\infty}\big\|_{\dot{E}} \leq C 2^{-i\varepsilon_-}\big\|\{\delta_{i,j} f_i\}_{j=-\infty}^{\infty}\big\|_{\dot{E}} \leq C 2^{-i\varepsilon_-}\|F\|_{\dot{E}},$$

and thus

$$\big\|\{\delta_{0,j}\textstyle\sum_{i=-\infty}^{0} 2^{iM}|f_i|\}_{j=-\infty}^{\infty}\big\|_{\dot{E}}^{\kappa} \leq C \sum_{i=-\infty}^{0} \big\|2^{iM}(S_-)^{-i}\{\delta_{i,j} f_i\}_{j=-\infty}^{\infty}\big\|_{\dot{E}}^{\kappa}$$
$$\leq C\|F\|_{\dot{E}}^{\kappa} \sum_{i=-\infty}^{0} 2^{i\kappa(M-\varepsilon_+)} = C\|F\|_{\dot{E}}^{\kappa}.$$

$\square$

COROLLARY 1.6.3. *Let $\dot{E} \in \dot{S}(\varepsilon_+, \varepsilon_-, r)$. Then for any $M > \varepsilon_-$ there is C so that*

$$\sum_{i=-\infty}^{0} 2^{iM} m_i \leq C\big\|\{m_i \chi(B(0, 2^{-i}))\}_{i=-\infty}^{\infty}\big\|_{\dot{E}}$$

for all sequences of nonnegative numbers $\{m_i\}_{i=-\infty}^{\infty}$ such that $m_i = 0$ for $i > 0$.

PROOF. By Lemma 1.6.2

$$\big\|\{\delta_{0,j}\chi(B(0,1))\}_{j=-\infty}^{\infty}\big\|_{\dot{E}} \sum_{i=-\infty}^{0} 2^{iM} m_i$$
$$\leq \big\|\{\delta_{0,j}\textstyle\sum_{i=-\infty}^{0} 2^{iM} m_i \chi(B(0, 2^{-i}))\}_{j=-\infty}^{\infty}\big\|_{\dot{E}}$$
$$\leq C\big\|\{m_j \chi(B(0, 2^{-j}))\}_{j=-\infty}^{\infty}\big\|_{\dot{E}}.$$

The corollary follows from the fact that $\big\|\{\delta_{0,j}\chi(B(0,1))\}_{j=-\infty}^{\infty}\big\|_{\dot{E}} > 0$. $\square$

COROLLARY 1.6.4. *If $\dot{E} \in \dot{S}(\varepsilon_+, \varepsilon_-, r)$, $F = \{f_i\}_{i=-\infty}^{\infty} \in \dot{E}$, and*

$$m_i = \inf_{x \in B(0, 2^{-i})} M_r f_i(x),$$

then for any $M > \varepsilon_-$

$$\sum_{i=-\infty}^{0} 2^{iM} m_i \leq C\|F\|_{\dot{E}}.$$

PROOF. By the previous corollary

$$\sum_{i=-\infty}^{0} 2^{iM} m_i \leq C\|\mathcal{M}_r F\|_{\dot{E}} \leq C\|F\|_{\dot{E}},$$

by Definition 1.6.1(b). □

COROLLARY 1.6.5. *Let $\dot{E} \in \dot{S}(\varepsilon_+, \varepsilon_-, r)$, and suppose that $F = \{f_i\}_{i=-\infty}^{\infty} \in \dot{E}$, satisfies $f_i \in \mathcal{S}'$, and $\operatorname{supp} \mathcal{F} f_i \subset B(0, 2^i)$. Then for any multi-index α with $|\alpha| > \varepsilon_-$, the series $\sum_{i=-\infty}^{0} |D^\alpha f_i(x)|$ converges uniformly on every compact set, and consequently the sum $\sum_{i=-\infty}^{0} D^\alpha f_i$ is an infinitely differentiable function.*

PROOF. By Lemma 1.1.7, for all $x \in B(0, 2^{-i})$

$$|D^\alpha f_i(x)| \leq C 2^{|\alpha|i} \inf_{y \in B(0, 2^{-i})} M_r f_i(y).$$

Thus for all x

$$\sum_{i=-\infty}^{0} |D^\alpha f_i(x)| \chi(B(0, 2^{-i}))(x) \leq C \sum_{i=-\infty}^{0} m_i 2^{|\alpha|i} \chi(B(0, 2^{-i}))(x).$$

The corollary follows from the previous corollary. □

DEFINITION 1.6.6. For any $M, L \in \mathbf{N}_0$ we denote by $\mathcal{S}'(M, L)$ the space of distributions $f \in \mathcal{D}'^L$ such that

(1.6.1) $$\sup_{1 \leq t < \infty} \sup_{\varphi} |\langle f, \varphi_t \rangle| t^{-M} < \infty,$$

where $\varphi_t = t^{-N} \varphi(\cdot/t)$, and $\sup_\varphi$ is taken over all φ that satisfy

(1.6.2) $$\varphi \in C_0^\infty(B(0, 1)), \quad \|\varphi\|_{C^L} \leq 1.$$

We denote by $\mathcal{S}'_0(M, L)$ the subspace of $\mathcal{S}'(M, L)$ consisting of such f that

$$\lim_{t \to \infty} \sup_{\varphi} |\langle f, \varphi_t \rangle| t^{-M} = 0,$$

where φ is as in (1.6.2).

REMARK. It is easily seen that $\mathcal{S}'(M, L) \subset \mathcal{S}'$. The condition (1.6.1) clearly defines a norm on $\mathcal{S}'(M, L)$.

DEFINITION 1.6.7. Let $r > 0$, and $M \in \mathbf{N}_0$. We denote by $\mathcal{L}_r(M)$ the set of functions $f \in L_{r,\mathrm{loc}}(\mathbf{R}^N)$ such that

$$\sup_{i \in \mathbf{N}_0} \|f\|_{L_r(B(0, 2^i))} 2^{-(M + \frac{N}{r})i} < \infty.$$

By $\mathcal{L}_{r,0}(M)$ we denote the set of functions $f \in \mathcal{L}_r(M)$, such that

$$\lim_{i \to \infty} \|f\|_{L_r(B(0, 2^i))} 2^{-(M + \frac{N}{r})i} = 0.$$

LEMMA 1.6.8. *Let $\dot{E} \in \dot{S}(\varepsilon_+, \varepsilon_-, r)$, and $F = \{f_i\}_{i=-\infty}^{\infty} \in \dot{E}$.*

(i) *Let $M \in \mathbf{N}_0$, $M > \varepsilon_-$, and suppose that $f_i \in \mathcal{S}' \cap \mathcal{L}_{r,0}(M)$, and $\operatorname{supp} \mathcal{F} f_i \subset B(0, 2^i)$. Then $\sum_{-\infty}^{0} f_i$ converges in the space $\mathcal{L}_r(M)/\mathfrak{P}_{M-1}$ to an element of $\mathcal{L}_{r,0}(M)/\mathfrak{P}_{M-1}$.*

(ii) *Let $M \in \mathbf{N}_0$ satisfy $M > \varepsilon_-$ and $M > N \max\{\frac{1}{r} - 1, 0\}$, and suppose that $f_i \in \mathcal{S}'$, and $\operatorname{supp} \mathcal{F} f_i \subset B(0, 2^{i+1}) \setminus B(0, 2^{i-1})$. Then, for any $L \in \mathbf{N}_0$ such that $L > N \max\{\frac{1}{r} - 1, 0\} - \varepsilon_+$, $\sum_{-\infty}^{0} f_i$ converges in the space $\mathcal{S}'(M, L)/\mathfrak{P}_{M-1}$ to an element in $\mathcal{S}'_0(M, L)/\mathfrak{P}_{M-1}$.*

PROOF. Set $\min_{|x| \leq 2^{-i}} M_r f_i(x) = m_i$. By Lemma 1.1.7 there is C so that

$$\max_{|x| \leq 2^{-i}} |D^\alpha f_i(x)| \leq C m_i 2^{i|\alpha|}, \quad |\alpha| \leq M.$$

Set $p_i(x) = \sum_{|\alpha| \leq M-1} D^\alpha f_i(0) x^\alpha / \alpha!$. Estimating the remainder term in Taylor's formula (cf. Corollary 1.1.8) gives

$$|f_i(x) - p_i(x)| \leq C m_i 2^{iM} |x|^M \quad \text{for } |x| \leq 2^{-i}.$$

Moreover, for $|x| \geq 2^{-i}$,

$$|p_i(x)| \leq C m_i \sum_{|\alpha| \leq M-1} 2^{i|\alpha|} |x|^{|\alpha|} \leq C m_i 2^{iM} |x|^M.$$

In order to prove part (i) of the lemma we observe that for $j \geq i$ we have

$$2^{jN} \int_{B(0, 2^{-j})} |f_i - p_i|^r \, dx \leq C(m_i 2^{(i-j)M})^r,$$

and for $j \leq i$

$$2^{jN} \int_{B(0, 2^{-j})} |p_i|^r \, dx \leq C(m_i 2^{(i-j)M})^r.$$

Moreover, for $j \leq i$,

$$2^{jN} \int_{B(0, 2^{-j})} |f_i|^r \, dx \leq C m_i^r \leq C(m_i 2^{(i-j)M})^r.$$

It follows that for all i,

$$\sup_j 2^{j(Mr+N)} \int_{B(0, 2^{-j})} |f_i - p_i|^r \, dx \leq C(m_i 2^{iM})^r.$$

Choose $\iota > 0$ so that $M - \iota > \varepsilon_-$. By Corollary 1.6.4 there is C so that

$$\sum_{i=-\infty}^{0} m_i 2^{i(M-\iota)} < C \|\{f_i\}\|_{\dot{E}}.$$

Let $\varepsilon > 0$ and choose n so large that $\sum_{i=-\infty}^{-n} m_i 2^{i(M-\iota)} < \varepsilon$. Assume that $0 < r < 1$ (the case $r \geq 1$ is simpler). It follows, using (1.2.4), that for any $m > n$

$$\sup_j 2^{j(M+\frac{N}{r})} \left\| \sum_{i=-n}^{0} (f_i - p_i) \right\|_{L_r(B(0, 2^{-j}))} \leq C \sum_{i=-m}^{-n} m_i 2^{i(M-\iota)} < C\varepsilon,$$

which proves convergence. If $f_i \in \mathcal{S}' \cap \mathcal{L}_{r,0}(M)$ for all i (which is always the case if $M > 0$), it follows easily that

$$\lim_{j \to -\infty} 2^{j(M+\frac{N}{r})} \left\| \sum_{i=-\infty}^{0} (f_i - p_i) \right\|_{L_r(B(0, 2^{-j}))} = 0,$$

which proves (i).

In order to prove (ii) we denote $\min_{|x|\leq 2^{-i+1}} M_r(M_r f_i)(x) = m_i^*$. Then $m_i^* \geq m_i$, but by Corollary 1.6.4 we still have

$$\sum_{i=-\infty}^{0} m_i^* 2^{iM} < C\|\{f_i\}\|_{\dot E}.$$

Let $\varphi \in C_0^\infty(B(0,1))$, $\|\varphi\|_{C^L} \leq 1$, and set $\varphi_t = t^{-N}\varphi(\,\cdot\,/t)$. As before

$$|f_i(x) - p_i(x)| \leq C m_i 2^{iM}|x|^M \quad \text{for } |x| \leq 2^{-i},$$

so that, for $t \leq 2^{-i}$,

$$|\langle f_i - p_i, \varphi_t\rangle| \leq \int_{\mathbf{R}^N} |f_i - p_i|\varphi_t\,dx \leq C t^M m_i 2^{iM}.$$

For $t \geq 2^{-i}$

$$\int_{\mathbf{R}^N} |p_i|\varphi_t\,dx \leq C t^M m_i 2^{iM},$$

and as in the proof of Proposition 1.1.12,

$$\int_{\mathbf{R}^N} |f_i|\varphi_t\,dx \leq C(t2^i)^{N\max\{1/r-1,0\}} \min_{|x|\leq 2^{-i+1}} M_r(M_r f_i)(x) \leq C t^M m_i^* 2^{iM},$$

if $M \geq N\max\{1/r-1, 0\}$.

It follows that for all $t \geq 1$, and for φ as in Definition 1.6.6

$$\sup_\varphi |\langle f_i - p_i, \varphi_t\rangle| t^{-M} \leq C m_i^* 2^{iM}.$$

Let $\varepsilon > 0$ and choose n so large that $\sum_{i=-\infty}^{-n} m_i^* 2^{iM} < \varepsilon$. Then for any $m > n$

$$\sup_{1\leq t<\infty}\sup_\varphi \big|\langle \sum_{i=-m}^{-n} f_i - \sum_{i=-m}^{-n} p_i, \varphi_t\rangle\big| t^{-M} \leq C \sum_{i=-m}^{-n} m_i^* 2^{iM} < C\varepsilon,$$

and the lemma follows. □

DEFINITION 1.6.9. Let $\dot E \in \dot S(\varepsilon_+, \varepsilon_-, r)$, where $\varepsilon_+, \varepsilon_- \in \mathbf{R}$, $r > 0$. The space $\dot Y(\dot E)$ consists of equivalence classes f of distributions, which admit a representation

(1.6.3) $$f = \sum_{i=-\infty}^{\infty} f_i,$$

where

(1.6.4) $$f_i \in \mathcal{S}', \quad \operatorname{supp} \mathcal{F} f_i \subset B(0, 2^{i+1}) \setminus B(0, 2^{i-1}),$$

(1.6.5) $$\|\{f_i\}_{i=-\infty}^\infty\|_{\dot E} < \infty,$$

and the series (1.6.3) converges in the sense of $\mathcal{S}'(M,L)/\mathfrak{P}_{M-1}$ for all $M, N \in \mathbf{N}_0$ satisfying $M > \varepsilon_-$, and $L > N\max\{\frac{1}{r}-1, 0\} - \varepsilon_+$, to $f \in \mathcal{S}_0'(M,L)/\mathfrak{P}_{M-1}$.

DEFINITION 1.6.10. Let $\dot E \in \dot S(\varepsilon_+, \varepsilon_-, r)$, where $\varepsilon_+, \varepsilon_-, r > 0$. The space $\dot{Y}L(\dot E)$ consists of equivalence classes f of functions, which admit a representation (1.6.3) satisfying (1.6.5), where

(1.6.6) $$f_i \in \mathcal{S}' \cap \mathcal{L}_{r,0}(M), \quad \operatorname{supp} \mathcal{F} f_i \subset B(0, 2^i),$$

and the series (1.6.3) converges in the sense of $\mathcal{L}_r(M)/\mathfrak{P}_{M-1}$ for all $M \in \mathbf{N}_0$ such that $M > \varepsilon_-$ to $f \in \mathcal{L}_{r,0}(M)/\mathfrak{P}_{M-1}$.

THEOREM 1.6.11. *Let $\dot{E} \in \dot{S}(\varepsilon_+, \varepsilon_-, r)$, where ε_+, ε_-, $r > 0$. Then, for any p such that $0 < p \leq \infty$, $\frac{N}{p} > \frac{N}{r} - \varepsilon_+$, and for any integer M such that $M > \varepsilon_-$, the following conditions on an element $f \in \mathcal{L}_{r,0}(M)/\mathfrak{P}_{M-1}$ are equivalent (with the usual modification if $p = \infty$):*
(i) $f \in \dot{Y}L(\dot{E})$.
(ii) *The estimate* (1.6.5) *is satisfied with*
$$f_i(x) = 2^{iN/p} \Big(\int_{B(x,2^{-i})} |\Delta_z^M f(x)|^p \, dz \Big)^{1/p}, \quad i \in \mathbf{Z}.$$
(iii) *The estimate* (1.6.5) *is satisfied with*
$$f_i(x) = 2^{iN/p} \mathcal{E}_M(f, B(x, 2^{-i}), L_p), \quad i \in \mathbf{Z}.$$
(iv) *f has a representation*
(1.6.7)
$$f(x) = \sum_{(i,k) \in \mathbf{Z} \times \mathbf{Z}^N} s_{i,k} a_{i,k}(x),$$
converging in the space $\mathcal{L}_{r,0}(M)/\mathfrak{P}_{M-1}$, such that the functions $b_{i,k}$ defined by $b_{i,k}(x) = a_{i,k}(2^{-i}(x-k))$ satisfy
(1.6.8)
$$b_{i,k} \in C_0^M([-1,2]^N), \quad \text{and} \quad \|b_{i,k}\|_{C^M} \leq 1,$$
and the coefficients $s_{i,k}$ are such that (1.6.5) *is satisfied with*
(1.6.9)
$$f_i = \sum_{k \in \mathbf{Z}^N} |s_{i,k}| \chi_{i,k}, \quad i \in \mathbf{Z}.$$

(v) *There are polynomials $\pi_{i,k} \in \mathfrak{P}_{M-1}$, $(i,k) \in \mathbf{Z} \times \mathbf{Z}^N$ such that* (1.6.5) *holds with*
$$f_i = \sum_{k \in \mathbf{Z}^N} |f - \pi_{i,k}| \chi_{Q_{i,k}(3)}, \quad i \in \mathbf{Z}.$$

THEOREM 1.6.12. *Let ε_+, $\varepsilon_- \in \mathbf{R}$, $r > 0$, and let $\dot{E} \in \dot{S}(\varepsilon_+, \varepsilon_-, r)$. Then, for any $M, L \in \mathbf{N}_0$ such that $M > \varepsilon_-$, and $L > N \max\{\frac{1}{r} - 1, 0\} - \varepsilon_+$, the following conditions on an element $f \in \mathcal{S}_0'(M,L)/\mathfrak{P}_{M-1}$ are equivalent:*
(i) $f \in \dot{Y}(\dot{E})$.
(ii) *The estimate* (1.6.5) *is satisfied with*
$$h_i(x) = 2^{iN} \sup_{\psi} |\langle f, \psi(2^i(\cdot - x)) \rangle|, \quad i \in \mathbf{Z},$$
where the supremum is taken over all $\psi \in \mathcal{S}$, satisfying (1.1.30)*, and such that $\psi \perp \mathfrak{P}_{M-1}$.*
(ii′) *The estimate* (1.6.5) *is satisfied with*
$$h_i(x) = 2^{iN} \sup_{\psi} |\langle f, \psi(2^i(\cdot - x)) \rangle|, \quad i \in \mathbf{Z},$$
where the supremum is taken over all $\psi \in C_0^\infty(B(0,1))$, such that $\|\psi\|_{C^L} \leq 1$, and $\psi \perp \mathfrak{P}_{M-1}$.
(iii) *f has a representation* (1.6.7)*, converging in $\mathcal{S}'(M,L)/\mathfrak{P}_{M-1}$, such that the functions $a_{i,k}$ in addition to* (1.6.8) *satisfy $a_{i,k} \perp \mathfrak{P}_{L-1}$, and the coefficients $s_{i,k}$ are such that* (1.6.5) *is satisfied with f_i defined by* (1.6.9)*.*

We prove Theorem 1.6.12 before proving Theorem 1.6.11.

1.7. Proof of Theorem 1.6.12

We need the following lemma.

LEMMA 1.7.1. *Let $f \in \mathcal{S}'_0(M,L)$, $M \in \mathbf{N}_0$, $L \in \mathbf{N}_0$, $\varphi \in \mathcal{S}$, $\varphi_t = t^{-N}\varphi(\cdot/t)$. Then $\lim_{t\to\infty} f * \varphi_t = 0$ in the space $\mathcal{S}'(M,L)/\mathfrak{P}_{M-1}$.*

PROOF. Let $\varepsilon > 0$. By assumption there is t_ε such that $|\langle f, \varphi_t\rangle| \leq \varepsilon t^M$ for $t \geq t_\varepsilon$. Here t_ε depends on $\sup_y(1+|y|^\lambda)|D^\gamma \varphi(y)|$ for $|\gamma| \leq L$ and a suitably large λ. It follows that there is C such that $|f * \varphi_t(x)| \leq C\varepsilon t^M$ for all $x \in B(0,t)$ and $t \geq t_\varepsilon$. Indeed, $f * \varphi_t(x) = \langle f, \varphi_t(x-\cdot)\rangle = \langle f, (\varphi(x-\cdot))_t\rangle$, and $\sup_y(1+|y|^\lambda)|D^\gamma \varphi(x-y)| \leq C\sup_y(1+|y|^\lambda)|D^\gamma \varphi(y)|$ for $|x| \leq 1$.

In the same way there is C, depending on φ and M, such that for all $x \in B(0,t)$,

$$|D^\alpha(f * \varphi_t)(x)| = |(f * D^\alpha \varphi_t)(x)| \leq C\varepsilon t^{M-|\alpha|}$$

for $0 \leq |\alpha| \leq M$.

Set $p_t(x) = \sum_{|\alpha| \leq M-1} D^\alpha(f*\varphi_t)(0)x^\alpha$, i.e., $p_t(x)$ is a Taylor polynomial of $f * \varphi_t(x)$. By the above estimates there is a constant C such that for any $t \geq t_\varepsilon$

$$|f*\varphi_t(x) - p_t(x)| \leq C\varepsilon|x|^M \quad \text{for } |x| \leq t.$$

Let $\psi \in C_0^\infty(B(0,1))$, $\psi_s = s^{-N}\psi(\cdot/s)$. Then for $s \leq t$

$$|\langle f*\varphi_t - p_t, \psi_s\rangle| \leq \int_{\mathbf{R}^N}|(f*\varphi_t - p_t)\psi_s|\,dx \leq C\varepsilon \int_{|x|\leq s}|x|^M|\psi_s|\,dx \leq C\varepsilon s^M.$$

For $s \geq t \geq t_\varepsilon$ we have

$$|\langle f*\varphi_t, \psi_s\rangle| = |\langle f, \varphi_t * \psi_s\rangle| \leq C\varepsilon s^M,$$

since $\varphi_t * \psi_s \in \mathcal{S}$, and $|D^\alpha(\varphi_t * \psi_s)(x)| \leq Cs^{-N-|\alpha|}(1+(|x|/s)^\lambda)$ for $|\alpha| \leq L$. Moreover,

$$|p_t(x)| \leq C\varepsilon \sum_{|\alpha|\leq M-1} s^{M-|\alpha|}|x|^\alpha,$$

so that $|\langle p_t, \psi_s\rangle| \leq C\varepsilon s^M$, since $\int_{\mathbf{R}^N}|x|^{|\alpha|}|\psi_s|\,dx = s^{|\alpha|}\int_{\mathbf{R}^N}|x|^{|\alpha|}|\psi|\,dx$. Thus, for any $t \geq t_\varepsilon$, and for ψ satisfying (1.6.2),

$$\sup_{s\geq 1}\sup_{\psi}|\langle f*\varphi_t - p_t, \psi_s\rangle|s^{-M} \leq C\varepsilon,$$

which proves the lemma. $\square$

COROLLARY 1.7.2. *Let $f \in \mathcal{S}'_0(M,L)$, $M \in \mathbf{N}_0$, $L \in \mathbf{N}_0$, $\varphi \in \mathcal{S}$, $\int_{\mathbf{R}^N}\varphi\,dx = 1$, and $\varphi_t = t^{-N}\varphi(\cdot/t)$. Then $f = \lim_{t\to 0} f*\varphi_t - \lim_{t\to\infty} f*\varphi_t$ in the space $\mathcal{S}'(M,L)/\mathfrak{P}_{M-1}$.*

PROOF. This follows from Lemma 1.7.1 and the well-known fact that in the sense of $\mathcal{S}'$, $f = \lim_{t\to 0} f*\varphi_t$. $\square$

PROOF OF THEOREM 1.6.12. The proof of Theorem 1.1.15 now requires only minor changes. We follow the same plan, and indicate the necessary modifications.

(ii) $\Rightarrow$ (i). Let $f \in \mathcal{S}'(M,L,0)/\mathfrak{P}_{M-1}$, let Ψ be defined as in (1.3.2), set $\Psi_j = 2^{jN}\Psi(\cdot 2^j)$ for $j \in \mathbf{Z}$, and $\psi_j = \Psi_j - \Psi_{j-1}$ for $j \in \mathbf{Z}$, so that $\sum_{j=-n+1}^m \psi_j = \Psi_m - \Psi_{-n}$. By Corollary 1.7.2 $\lim_{m\to\infty} f*\Psi_m = f$ in $\mathcal{S}'$, and $\lim_{n\to\infty} f*\Psi_{-n} = 0$ in $\mathcal{S}'(M,L)/\mathfrak{P}_{M-1}$. Thus, setting $f_j = f*\psi_j$, we have $f = \sum_{j=-\infty}^\infty f_j$, with convergence in $\mathcal{S}'(M,L)/\mathfrak{P}_{M-1}$, and $\operatorname{supp}\mathcal{F}f_j \subset \operatorname{supp}\mathcal{F}\psi_j \subset B(0,2^{j+1})\setminus B(0,2^{j-1})$.

Now assume that f satisfies the conditions in (ii). For any $\lambda \geq 0$ and any positive integer L there is a constant C such that
$$\max_x (1+|x|^\lambda)|D^\alpha \Psi(x)| \leq C \quad \text{for all } \alpha \text{ with } |\alpha| \leq L .$$
Since all moments of ψ_j vanish for $j \geq 1$, it follows from (1.1.30) that for the same constant C, the functions $f_j(x) = f * \psi_j(x) = \langle f, \psi_j(x-\cdot)\rangle$ satisfy $|f_j| \leq Ch_j$, with h_j defined as in (ii). The claim follows.

(iii) $\Rightarrow$ (ii). Suppose that $f \in \mathcal{S}_0'(M, L)/\mathfrak{P}_{M-1}$, and that f has a representation (1.6.7),
$$f = \sum_{j=-\infty}^\infty f_j, \quad f_j = \sum_{k \in \mathbf{Z}^N} s_{j,k} a_{j,k},$$
where $\lim_{n\to\infty} \sum_{j=-n}^\infty f_j = f$ with convergence in the sense of $\mathcal{S}'(M, L)/\mathfrak{P}_{M-1}$. Then $\langle f, \psi\rangle = \sum_{j=-\infty}^\infty \langle f_j, \psi\rangle$ for any $\psi \in \mathcal{S}$ such that $\psi \perp \mathfrak{P}_{M-1}$, and the claim follows as in the proof of Theorem 1.1.15.

(ii') $\Rightarrow$ (iii) and (i) $\Rightarrow$ (iii). These implications follow as before. $\square$

1.8. Proof of Theorem 1.6.11

We follow the same plan as in the proof of Theorem 1.1.14, and indicate the necessary changes. Again, we let $q = \min\{p, r\}$, and denote by (ii*) and (iii*) the conditions (ii) and (iii) with p replaced by q.

We need the following lemma, which plays the same role as Lemma 1.7.1, and complements Lemma 1.2.3.

LEMMA 1.8.1. *Let $f \in \mathcal{L}_{q,0}(M)/\mathfrak{P}_{M-1}$, and suppose that the assumption (iii*) is satisfied. Let $\omega \in C_0^\infty(Q_{0,0}(2))$ satisfy $\sum_{k \in \mathbf{Z}^N} \omega(x-k) = 1$, and set $\omega_{i,k}(x) = \omega(2^i x - k)$. Set*
$$f_i(x) = \sum_{k \in \mathbf{Z}^N} \pi_{i,k}(x)\omega_{i,k}(x), \quad i \in \mathbf{Z},$$
where $\pi_{i,k} \in \mathfrak{P}_{M-1}$ are polynomials such that
$$\|f - \pi_{i,k}\|_{L_q(Q_{i,k}(3))} \leq 2\mathcal{E}_m(f, Q_{i,k}(3), L_q).$$
Then $\lim_{i\to-\infty} f_i = 0$ in $\mathcal{L}_r(M)/\mathfrak{P}_{M-1}$.

PROOF. Choose $\varepsilon > 0$, let $-i$ be so large that
$$\|f\|_{L_q(B(0,2^{-i}))} < \varepsilon 2^{-(N/r+M)i},$$
and set $p_i(x) = \sum_{|\alpha| \leq M-1} D^\alpha f_i(0) x^\alpha / \alpha!$.

There is C such that $\max_{|y| \leq 1} |D^\alpha \pi_{0,0}(y)| \leq C\|\pi_{0,0}\|_{L_r(Q_{0,0})} \leq C\|f\|_{L_q(Q_{0,0})}$ for all $|\alpha| \leq M$. By a change of scale it follows that
$$|D^\alpha(\pi_{i,k}\omega_{i,k})(y)| \leq C \sum_{|\beta| \leq |\alpha|} |D^\beta \pi_{i,k}(y) D^{\alpha-\beta}\omega_{i,k}(y)| \leq C 2^{i|\alpha|} 2^{iN/r} \|f\|_{L_q(Q_{i,k})}$$
for $|\alpha| \leq M$ and $|y| \leq 2^{-i}$. It follows that for a certain i_0 depending on N
$$\max_{|\alpha|=M} \max_{|y| \leq 2^{-i}} |D^\alpha f_i(y)| \leq C 2^{i(N/r+M)} \|f\|_{L_q(B(0,2^{-i+i_0}))} \leq C\varepsilon,$$
and thus for $x \in B(0, 2^{-i})$ we have
$$|f_i(x) - p_i(x)| \leq C\varepsilon |x|^M,$$

and for $j \geq i$
$$\|f_i - p_i\|_{L_q(B(0,2^{-j}))} \leq C\varepsilon 2^{-(N/r+M)j}.$$
Moreover,
$$|p_i(x)| \leq C \sum_{|\alpha| \leq M-1} |D^\alpha f_i(0)||x|^{|\alpha|}$$
$$\leq C \sum_{|\alpha| \leq M-1} 2^{i|\alpha|} 2^{iN/r} |x|^{|\alpha|} \|f\|_{L_q(B(0,2^{-i+i_0}))} \leq C\varepsilon 2^{-iM} \sum_{|\alpha| \leq M-1} 2^{i|\alpha|} |x|^{|\alpha|},$$
and thus, for $j < i$,
$$2^{jN/q}\|p_i\|_{L_q(B(0,2^{-j}))} \leq C\varepsilon 2^{-iM} \sum_{|\alpha| \leq M} 2^{(i-j)|\alpha|} \leq C\varepsilon 2^{-jM}.$$
Finally, for $j < i$,
$$2^{jN} \int_{B(0,2^{-j})} |f_i|^q \, dx = 2^{jN} \int_{B(0,2^{-j})} |\sum_k \pi_{i,k} \omega_{i,k}|^q \, dx$$
$$\leq C 2^{jN} \sum_{Q_{i,k}(3) \cap B(0,2^{-j}) \neq \emptyset} \int_{Q_{i,k}(3)} |\pi_{i,k}|^q \, dx$$
$$\leq C 2^{jN} \int_{B(0,2^{-j+i_0})} |f|^q \, dx \leq C\varepsilon 2^{-jMq},$$
which proves the lemma. $\square$

PROOF OF THEOREM 1.6.11. (i) $\Rightarrow$ (iv). This implication follows as before.
(iv) $\Rightarrow$ (i). Suppose that $f \in \mathcal{L}_{q,0}(M)/\mathfrak{P}_{M-1}$, and that f has a representation
$$f(x) = \sum_{i=-\infty}^{\infty} h_i(x), \quad h_i(x) = \sum_{k \in \mathbf{Z}^N} s_{i,k} a_{i,k}(x)$$
that satisfies the conditions in (iv). Define Ψ as in (1.3.2), $\Psi_{2^{-j}} = 2^{jN}\Psi(\cdot\, 2^j)$, and $\psi_j = \Psi_{2^{-j}} - \Psi_{2^{-(j-1)}}$ for $j \in \mathbf{Z}$, and set $f_{i,j} = h_i * \psi_j$ for $j > i$, and $f_{i,i} = h_i * \Psi_{2^{-j}}$ as before. Then
$$f = \sum_{i=-\infty}^{\infty} \sum_{j=i}^{\infty} f_{i,j} = \sum_{l=0}^{\infty} \sum_{j=-\infty}^{\infty} f_{j-l,j}.$$
It follows as before that for each l
$$\|\{f_{j-l,j}\}_{j=-\infty}^{\infty}\|_{\dot{E}} \leq C 2^{-l(\varepsilon_- + \iota)} \|\{M_{r,d}g_{j-l}\}_{j=-\infty}^{\infty}\|_{\dot{E}}$$
$$= C 2^{-l(\varepsilon_- + \iota)} \|S_-^l \{M_{r,d}g_j\}_{j=-\infty}^{\infty}\|_{\dot{E}} \leq C 2^{-l\iota} \|\{g_j\}_{j=-\infty}^{\infty}\|_{\dot{E}},$$
and hence each $\sum_{j=-\infty}^{\infty} f_{j-l,j}$ represents an element $F_l \in \dot{Y}L(\dot{E})$ with
$$\|F_l\|_{\dot{Y}L(\dot{E})} \leq C 2^{-l\iota} \|\{g_i\}_{i=-\infty}^{\infty}\|_{\dot{E}}.$$
It follows that $f = \sum_{l=0}^{\infty} F_l$ belongs to $\dot{Y}L(\dot{E})$ and
$$\|f\|_{\dot{Y}L(\dot{E})} \leq C\|\{g_i\}_{i=-\infty}^{\infty}\|_{\dot{E}}.$$
(ii*) $\Rightarrow$ (v). The implication follows from Lemma 1.2.1 as before.
(v) $\Rightarrow$ (iii*). The implication follows as before.

(iii*) ⇒ (iv). As in the proof of the corresponding implication in Theorem 1.1.14 we set
$$f_i - f_{i-1} = \sum_{k \in \mathbf{Z}^N} s_{i,k} a_{i,k},$$
and by Lemma 1.2.3
$$f - f_n = \sum_{i=n+1}^{\infty} \sum_{k \in \mathbf{Z}^N} s_{i,k} a_{i,k}$$
with convergence in $L_{r,\mathrm{loc}}$ and almost everywhere. By the above we have
$$f = \lim_{n \to -\infty} \sum_{i=n}^{\infty} \sum_{k \in \mathbf{Z}^N} s_{i,k} a_{i,k}$$
with convergence in the sense of $\mathcal{L}_{q,0}(M)/\mathfrak{P}_{M-1}$.

(iv) ⇒ (ii). We assume $f \in \mathcal{L}_{q,0}(M)/\mathfrak{P}_{M-1}$ and that f has a representation
$$f(x) = \sum_{i=-\infty}^{\infty} h_i(x), \quad h_i(x) = \sum_{k \in \mathbf{Z}^N} s_{i,k} a_{i,k}(x)$$
that satisfies the conditions in (iv). We now have to use the estimate $|\Delta_z^m h_j(x)| \leq C 2^{-im} 2^{jm} \sum_l |s_{j,l}|$ for all j, $-\infty < j < i$, but the desired estimate follows as before from Lemma 1.2.6, which is valid almost without change.

(iv) ⇒ (iii). This implication follows as the previous one after the same modification of Lemma 1.2.6. □

CHAPTER 2

Differentiability and Spectral Synthesis

2.1. Capacities and Differentials

DEFINITION 2.1.1. Let $s \geq 0$, $\varepsilon_+, \varepsilon_- \in \mathbf{R}$, $0 < r \leq \infty$, let $E \in S(\varepsilon_+, \varepsilon_-, r)$, and let $A \subset \mathbf{R}^N$. We define the s-capacity of A associated to the space E by
$$\mathrm{cap}(A, s, E) = \inf \|\{g_i\}_{i=0}^\infty\|_E,$$
where the infimum is taken over all function sequences $\{g_i\}_{i=0}^\infty$ with
$$g_i(x) = \sum_{k \in \mathbf{Z}^N} s_{i,k} \chi_{i,k}(x), \quad s_{i,k} \geq 0,$$
such that for all $x \in A$
$$\sum_{i=0}^\infty 2^{is} g_i(x) \geq 1.$$
A property is said to hold *quasieverywhere* on a set G with respect to $\mathrm{cap}(\,\cdot\,, s, E)$ (abbreviated $\mathrm{cap}(\,\cdot\,, 0, E)$-q.e.), if there is a set A such that $\mathrm{cap}(A, s, E) = 0$ and the property holds at all $x \in G \setminus A$.

REMARK. If $E = L_p(l_2^\lambda)$ with $1 < p < \infty$, $0 \leq s < \lambda$, in which case $YL(E) = Y(E) = F_{p,2}^\lambda = L_p^\lambda$, then $\mathrm{cap}(\,\cdot\,, s, E)^p$ is equivalent to the capacity denoted $C_{\lambda-s,p}(\cdot)$ in e.g. [4]. See Definition 2.3.3, and also Definitions 4.4.2 and 4.7.1, and Proposition 4.7.2 in [4]. References to the literature on capacities associated to B- and F-spaces for $p > 1$ are found in [4], Section 5.6.7. For $0 < p \leq 1$, see D. R. Adams [1], [2], [3], and Netrusov [33], [34], [36].

LEMMA 2.1.2. *With the notation of Definition 2.1.1*
$$\mathrm{cap}(A, s, E) = 0$$
if and only if there is there is a function sequence $\{g_i\}_{i=0}^\infty$ with
$$g_i(x) = \sum_{k \in \mathbf{Z}^N} s_{i,k} \chi_{i,k}(x), \quad s_{i,k} \geq 0,$$
and $\|\{g_i\}_{i=0}^\infty\|_E < \infty$, such that for all $x \in A$
$$\sum_{i=0}^\infty 2^{is} g_i(x) = \infty.$$

PROOF. Suppose that $\mathrm{cap}(A, s, E) = 0$. For each $n \in \mathbf{N}$ there is a sequence $\{g_i^{(n)}\}_{i=0}^\infty$ with $g_i^{(n)} = \sum_{k \in \mathbf{Z}^N} s_{i,k}^{(n)} \chi_{i,k}$, $s_{i,k}^{(n)} \geq 0$, such that $\sum_{i=0}^\infty 2^{is} g_i^{(n)}(x) \geq 1$ on A and $\|\{g_i^{(n)}\}_{i=0}^\infty\|_E < 2^{-n}$. Set $g_i = \sum_{n=0}^\infty g_i^{(n)}$. Then $\sum_{i=0}^\infty 2^{is} g_i(x) = \infty$ on A, and, by (1.1.2), $\|\{g_i\}_{i=0}^\infty\|_E^\kappa \leq C \sum_{n=0}^\infty 2^{-n\kappa} < \infty$.

The converse is obvious. $\square$

DEFINITION 2.1.3. If $f \in L_{r,\text{loc}}$, $r > 0$, we say that a polynomial π_s of degree at most s is an L^r-*differential of order* s to f at a point x if
$$\lim_{\rho \to 0} \rho^{-N-sr} \int_{B(x,\rho)} |f(y) - \pi_s(y)|^r \, dy = 0.$$

It is easily seen that such a polynomial is unique, and thus, if $f \in C^s$, the differential is the Taylor polynomial of order s,
$$P_x^s f(y) = \sum_{|\beta| \leq s} \frac{D^\beta f(x)}{\beta!} (y - x)^\beta.$$

To say that f has a differential of order 0 at x is just another way of saying that x is a Lebesgue point for f.

By Theorem 1.1.14 a function f in $YL(E)$, where $E \in S(\varepsilon_+, \varepsilon_-, r)$, can be represented as

(2.1.1) $\qquad f = \sum_{i=0}^{\infty} f_i, \text{ where } f_i = \sum_{k \in \mathbf{Z}^N} s_{i,k} a_{i,k}, \text{ and } \{f_i\}_{i=0}^\infty \in E.$

Here the $a_{i,k}$ are such that if $M > \varepsilon_-$ and g_i is defined by

(2.1.2) $\qquad g_i(x) = \sum_{k \in \mathbf{Z}^N} s_{i,k} \chi_{i,k}(x),$

where

(2.1.3) $\qquad s_{i,k} = \max \{ 2^{-i|\beta|} |D^\beta f_i(x)| : x \in Q_{i,k}(3), |\beta| \leq M \},$

then

(2.1.4) $\qquad \|\{g_i\}_{i=0}^\infty\|_E \leq C \|f\|_{YL(E)},$

with a C independent of f.

Lemma 2.1.2 gives that

(2.1.5) $\qquad \sum_{i=0}^{\infty} |f_i(x)| < \infty \quad \text{cap}(\,\cdot\,, 0, E)\text{-q.e.}$

Similarly, for any multi-index β,

(2.1.6) $\qquad \sum_{i=0}^{\infty} |D^\beta f_i(x)| < \infty \quad \text{cap}(\,\cdot\,, |\beta|, E)\text{-q.e.}$

It follows that for any s, the Taylor polynomial $P_x^s f$ is well defined for x off a set with zero $(\,\cdot\,, s, \dot{E})$-capacity.

The following theorem makes these ideas much more precise. In the case of BL-spaces the result is found in Netrusov [34], Assertion 1.2 and Lemma 1.3. The theorem is proved in [4], Section 10.1, for the spaces $F_{p,2}^\lambda = L_p^\lambda$, $1 < p < \infty$, and the general proof is similar, but in view of its critical importance for the applications given in the following sections, we give the full proof here.

THEOREM 2.1.4. *Let $f \in YL(E)$, where $E \in S(\varepsilon_+, \varepsilon_-, r)$, $\varepsilon_+, \varepsilon_- > 0$, and $r > 0$. Suppose that $\{f_i\}_{i=0}^\infty$ and $\{g_i\}_{i=0}^\infty$ are functions such that the conditions (2.1.1) – (2.1.4) are satisfied. Let s be an integer, $s \leq \varepsilon_-$, and let $P_x^s f_i$ be the Taylor polynomial of degree s of f_i at x. Then there is a set F_s with $\text{cap}(F_s, s, E) = 0$, such that for every $x \in F_s^c$:*

(i) $\sum_{i=0}^{\infty} |D^\beta f_i(x)| < \infty \quad \text{for } |\beta| \leq s$;
(ii) f has an L^r-differential of order s at x;
(iii) The differential is $P_x^s f(y) = \sum_{|\beta| \leq s} D^\beta f(x)(y-x)^\beta/\beta!$, where $D^\beta f(x) = \sum_{i=0}^{\infty} D^\beta f_i(x)$ for $|\beta| \leq s$.

In what follows, when we write $f(x)$ for an $f \in YL(E)$, we will always tacitly assume that x is a Lebesgue point for f, and that $f(x)$ is given by

$$\lim_{\rho \to 0} \rho^{-N} \int_{B(x,\rho)} |f(y) - f(x)|^r \, dy = 0. \tag{2.1.7}$$

The proof of the theorem depends on the following basic lemma.

LEMMA 2.1.5. *Let $g_i(x) = \sum_{k \in \mathbf{Z}^N} s_{i,k} \chi_{i,k}(x), s_{i,k} \geq 0$, for $i \in \mathbf{N}_0$, and suppose that $0 < \|\{g_i\}_{i=0}^\infty\|_E < \infty$. Then there exist h_i, $i \in \mathbf{N}_0$, with the following properties:*

(i) $h_i(x) = \sum_{k \in \mathbf{Z}^N} t_{i,k} \chi_{i,k}(x), \quad t_{i,k} \geq 0$;
(ii) $g_i \leq h_i$;
(iii) *For all $Q_{i,k}$*

$$\left(2^{iN} \int_{Q_{i,k}(3)} \left(\sum_{j=i}^\infty g_j\right)^r dx\right)^{1/r} \leq t_{i,k};$$

(iv) *There is a constant $C > 0$ independent of $\{g_i\}_{i=0}^\infty$ such that*

$$\|\{h_i\}_{i=0}^\infty\|_E \leq C\|\{g_i\}_{i=0}^\infty\|_E;$$

(v) $\sum_{i=0}^\infty 2^{iM} h_i(x) = \infty$ *for all x if $M > \varepsilon_-$*;
(vi) *There is a constant $C > 0$ such that $t_{i,k} \leq C t_{i,k'}$ if $|k - k'| = 1$, $i \in \mathbf{N}_0$.*

PROOF. Set

$$t_{i,k} = \left(2^{iN} \int_{Q_{i,k}(3)} \left(\sum_{j=i}^\infty g_j\right)^r dx\right)^{1/r}.$$

Then (ii) and (iii) are trivially satisfied, and (iv) is satisfied by Lemma 1.2.4.

If now (v) is not satisfied, we can modify the sequence $\{t_{i,k}\}$ by adding suitable functions to h_i. We let $M > \varepsilon_-$, and we define

$$h_i'(x) = h_i(x) + \delta 2^{-iM} \sum_{k \in \mathbf{Z}^n} \frac{\chi_{0,0}(x-k)}{(1+|k|)^\lambda},$$

where δ is a small positive number, and λ is so large that $\sum_{k \in \mathbf{Z}^n} \frac{\chi_{0,0}(x-k)}{(1+|k|)^\lambda} \leq C M_{r,d} \chi_{0,0}(x)$.

If δ is chosen sufficiently small, the modified sequence $\{h_i'\}_0^\infty$ satisfies conditions (i) – (v).

In order to satisfy (vi), we write the function already constructed as

$$h_i' = \sum_{k \in \mathbf{Z}^n} t_{i,k}' \chi_{i,k},$$

and we define a modified function h_i'' by setting

$$h_i'' = \sum_{k \in \mathbf{Z}^n} t_{i,k}' u_{i,k} = \sum_{k \in \mathbf{Z}^n} t_{i,k}'' \chi_{i,k},$$

where

$$u_{i,k}(x) = u_0(2^i x - k), \quad \text{and} \quad u_0(x) = \sum_{k \in \mathbf{Z}^n} \frac{\chi_{0,0}(x-k)}{(1+|k|)^\lambda},$$

with the same λ as above. It is then easy to see that h_i'' satisfies (i) – (iii), and (v). That (vi) is satisfied follows from the construction, and the trivial fact that $(1+|k|)(1+|k'|)^{-1}$ is bounded above and below for $|k - k'| \le 1$.

Finally, it follows from Lemma 1.2.7 (with $i = j$) that

$$h_i'' = \sum_{k \in \mathbf{Z}^n} t_{i,k}' u_{i,k} \le CM_{r,d} h_i',$$

which proves (iv) by Definition 1.1.1(b). $\square$

PROOF OF THEOREM 2.1.4. By assumption $|D^\beta f_i| \le C\, 2^{i|\beta|} g_i$. Let $\{h_i\}_{i=0}^\infty$ be the functions constructed in Lemma 2.1.5. Set

(2.1.8) $\qquad F_s = \{\, x : \sum_{i=0}^\infty 2^{is} h_i(x) = \infty\, \}, \quad s = 0, 1, \ldots.$

Then $\operatorname{cap}(F_s, s, E) = 0$ by Lemma 2.1.2, but we note that $F_s = \mathbf{R}^N$ for $s > \varepsilon_-$ by (v) in Lemma 2.1.5. Let $s \le \varepsilon_-$. By (2.1.3),

$$|D^\beta f_i(x)| \le C 2^{i|\beta|} g_i(x) \quad \text{for } |\beta| \le s$$

so that

$$\sum_{i=0}^\infty |D^\beta f_i(x)| \le \sum_{i=0}^\infty 2^{i|\beta|} g_i(x) < \infty$$

for all $x \in F_s^c$ and $|\beta| \le s$. Thus, we can define $P_x^s f$ by

$$P_x^s f(y) = \sum_{i=0}^\infty P_x^s f_i(y) = \sum_{|\beta| \le s} \frac{(y-x)^\beta}{\beta!} \sum_{i=0}^\infty D^\beta f_i(x), \quad x \in F_s^c.$$

We claim that

$$\lim_{j \to \infty} 2^{j(N+sr)} \int_{B(x, 2^{-j})} |f(y) - P_x^s f(y)|^r\, dy = 0, \quad x \in F_s^c.$$

In fact, for $i \le j$, and $y \in B(x, 2^{-j})$,

$$|f_i(y) - P_x^s f_i(y)| \le C\, 2^{-j(s+1)} \sum_{|\beta|=s+1} \max_{z \in B(x, 2^{-j})} |D^\beta f_i(z)| \le C 2^{(i-j)(s+1)} g_i(x),$$

and thus,

$$2^{j(N+sr)} \int_{B(x, 2^{-j})} \Big|\sum_{i=0}^j f_i - \sum_{i=0}^j P_x^s f_i\Big|^r dy \le C \Big(\sum_{i=0}^j 2^{i(s+1)-j} g_i(x)\Big)^r.$$

Further, by Lemma 2.1.5,

$$2^{j(N+sr)} \int_{B(x, 2^{-j})} \Big(\sum_{i=j+1}^\infty |f_i|\Big)^r dy \le C 2^{(j+1)sr} h_{j+1}(x)^r,$$

and by (2.1.3)

$$2^{j(N+sr)} \int_{B(x, 2^{-j})} \Big(\sum_{i=j+1}^\infty |P_x^s f_i|\Big)^r dy \le C \Big(\sum_{i=j+1}^\infty 2^{is} g_i(x)\Big)^r.$$

Thus
$$2^{j(N+sr)} \int_{B(x,2^{-j})} \Big|\sum_{i=0}^{\infty} f_i - \sum_{i=0}^{\infty} P_x^s f_i\Big|^r dy$$
$$\leq C\Big(\sum_{i=0}^{j} 2^{i-j} 2^{is} g_i(x)\Big)^r + C 2^{(j+1)sr} h_{j+1}(x)^r + C\Big(\sum_{i=j+1}^{\infty} 2^{is} g_i(x)\Big)^r.$$

The theorem now follows from the convergence of $\sum_{i=0}^{\infty} 2^{is} g_i(x)$ and $\sum_{i=0}^{\infty} 2^{is} h_i(x)$, and the elementary fact that
$$\lim_{j \to \infty} \sum_{i=0}^{j} 2^{i-j} a_i = 0,$$
if $a_i \geq 0$ and $\sum_{i=0}^{\infty} a_i < \infty$. □

The following corollary will be crucial in proving Theorem 2.2.1 below.

COROLLARY 2.1.6. *Let ε_+, ε_-, $r > 0$, let s be an integer, $0 \leq s \leq \varepsilon_-$. Let $E \in S(\varepsilon_+, \varepsilon_-, r)$, and let $f \in YL(E)$. Suppose that there is a set $A \subset \mathbf{R}^N$ such that f satisfies*
$$\lim_{\rho \to 0} \rho^{-N-sr} \int_{B(x,\rho)} |f(y)|^r dy = 0$$
quasieverywhere on A with respect to $\mathrm{cap}(\,\cdot\,, s, E)$ for all integers $s \leq \varepsilon_-$. Then there exist functions $\{f_i\}_{i=0}^{\infty}$ and $\{g_i\}_{i=0}^{\infty}$ satisfying (2.1.1) – (2.1.4), and functions $\{h_i\}_{i=0}^{\infty}$ satisfying Lemma 2.1.5, such that if $x \in A$, and $\sum_{i=0}^{\infty} 2^{is} h_i(x) < \infty$ for some integer s, then $\sum_{i=0}^{\infty} |D^\beta f_i(x)| < \infty$, and

(2.1.9)
$$\sum_{i=0}^{\infty} D^\beta f_i(x) = 0$$
for all β, $|\beta| \leq s$.

PROOF. Let $\{f_i\}_{i=0}^{\infty}$ and $\{h_i\}_{i=0}^{\infty}$ satisfy the conditions in Lemma 2.1.5. Let $A_s = \{x \in A : P_x^s f \neq 0\}$, $0 \leq s \leq \varepsilon_-$. Then, by Lemma 2.1.2, for any $\varepsilon > 0$ there exist functions $h'_{s,i}(x)$ such that
$$h'_{s,i} = \sum_{k \in \mathbf{Z}^N} t'_{s,i,k} \chi_{i,k}, \quad t'_{s,i,k} \geq 0,$$
$$\|\{2^{i\alpha} h'_{s,i}\}_{i=0}^{\infty}\|_E \leq \varepsilon \|f\|_{YL(E)},$$
and
$$\sum_{i=0}^{\infty} 2^{is} h'_{s,i}(x) = \infty \quad \text{for all } x \in E_s.$$

Set
$$h'_i = \sum_{0 \leq s \leq \varepsilon_-} h'_{s,i} = \sum_{k \in \mathbf{Z}^N} t'_{i,k} \chi_{i,k}.$$
We define functions h''_i by setting
$$h''_i = h_i + \sum_{k \in \mathbf{Z}^N} t'_{i,k} u_{i,k} = \sum_{k \in \mathbf{Z}^N} t''_{i,k} \chi_{i,k},$$
where $u_{i,k}$ is defined as in the proof of Lemma 2.1.5. Then h''_i satisfies all required conditions. □

2.2. Spectral Synthesis

The main result of this section is the following so called *spectral synthesis* theorem for spaces $YL(E)$. See [**24**], and [**4**], Sections 9.1 and 9.13 for background and history of the problem and for a discussion of the name "spectral synthesis". Under somewhat less general hypotheses the theorem was announced in [**35**] and (under still less general assumptions) proved in [**4**], Section 10.1. A spectral synthesis result which is not covered by this general theorem is found in [**37**].

The proof given here is basically the same as that given in [**4**], but for the reader's convenience we give it in full detail here, at the cost of some repetition.

We will need to assume that, in addition to belonging to some class $S(\varepsilon_+, \varepsilon_-, r)$, the spaces E will satisfy the following condition:

If $g = \{g_i\}_{i=0}^{\infty} \in E$ with $g_i(x) = \sum_{k \in \mathbf{Z}^N} s_{i,k} \chi_{i,k}(x)$, and if $g^{(n)} = \{g_i^{(n)}\}_{i=0}^{\infty}$ is defined by $g_i^{(n)} = 0$ for $i \leq n$ and $g_i^{(n)} = g_i$ for $i > n$, then

$$(2.2.1) \qquad \lim_{n \to \infty} \|\{g_i^{(n)}\}_{i=0}^{\infty}\|_E = 0.$$

THEOREM 2.2.1. *Let ε_+, ε_-, $r > 0$, let $E \in S(\varepsilon_+, \varepsilon_-, r)$, and assume furthermore that E satisfies the condition* (2.2.1). *Let $A \subset \mathbf{R}^N$, and suppose that $f \in YL(E)$. Then the following conditions are equivalent:*

(i) *For all integers s, $0 \leq s \leq \varepsilon_-$,*

$$(2.2.2) \qquad \lim_{\rho \to 0} \rho^{-N-sr} \int_{B(x,\rho)} |f(y)|^r \, dy = 0$$

quasieverywhere on A with respect to $\mathrm{cap}(\,\cdot\,, s, E)$.

(ii) *For all $\varepsilon > 0$ there exists $f_\varepsilon \in YL(E)$ such that $\mathrm{supp}\, f_\varepsilon \cap A = \emptyset$, and $\|f - f_\varepsilon\|_{YL(E)} < \varepsilon$.*

(iii) *For all $\varepsilon > 0$ there exists η_ε such that $\mathrm{supp}\, \eta_\varepsilon \cap A = \emptyset$, $0 \leq \eta_\varepsilon \leq 1$, and $\|f - f\eta_\varepsilon\|_{YL(E)} < \varepsilon$.*

REMARK. The condition (2.2.1) implies that functions with compactly supported Fourier transform are dense in $YL(E)$. It is typically not satisfied if $E = l_\infty^\lambda(L_p)$, which corresponds to the Besov space $B_{\infty,p}^\lambda$. (See the examples following Definition 1.1.1.) On the other hand (2.2.1) is satisfied if we replace l_∞^λ by the corresponding space c_0^λ of sequences converging weightedly to 0, which gives a space which we denote by $B_{\infty,p,0}^\lambda$.

The following elementary lemma will play an important role in the construction of the multiplier η_ε in Theorem 2.2.1(iii). Se [**4**], 10.4, p. 303, for remarks on the interesting history of this lemma.

LEMMA 2.2.2. *Let $a_i > 0$, $i = 0, 1, 2, \ldots$, and let $s \geq 0$.*

(i) *If $\sum_{i=0}^{\infty} a_i = \infty$, then*

$$\sum_{i=0}^{\infty} \frac{a_i}{A_i} = \infty,$$

where $A_i = \sum_{j=0}^{i} a_j$.

(ii) *If $\sum_{i=0}^{\infty} a_i < \infty$, then*

$$\sum_{i=0}^{\infty} \frac{a_i}{B_i} = \infty,$$

where $B_i = \sum_{j=i}^{\infty} a_j$.

(iii) If $\sum_{i=0}^{\infty} a_i 2^{is} < \infty$, but $\sum_{i=0}^{\infty} a_i 2^{i(s+1)} = \infty$, then
$$\sum_{i=0}^{\infty} \frac{a_i 2^{is}}{2^{-i} A_i + B_i} = \infty,$$
where $A_i = \sum_{j=0}^{i} a_j 2^{j(s+1)}$, and $B_i = \sum_{j=i+1}^{\infty} a_j 2^{js}$.

PROOF. Parts (i) and (ii) are well-known. To prove (i) we observe that if $\lim_{n\to\infty} A_n = \infty$, then for any k
$$\liminf_{n\to\infty} \sum_{i=k}^{n} \frac{a_i}{A_i} \geq \liminf_{n\to\infty} \frac{1}{A_n} \sum_{i=k}^{n} a_i = \liminf_{n\to\infty} \frac{A_n - A_{k-1}}{A_n} = 1.$$

In (ii) similarly, if all $B_n > 0$ and $\lim_{n\to\infty} B_n = 0$, then for any k
$$\liminf_{n\to\infty} \sum_{i=k}^{n} \frac{a_i}{B_i} \geq \liminf_{n\to\infty} \frac{1}{B_k} \sum_{i=k}^{n} a_i = \liminf_{n\to\infty} \frac{B_k - B_{n+1}}{B_k} = 1.$$

In proving (iii) we consider two cases: we either have
$$\liminf_{n\to\infty} 2^{-n} A_n / B_n \geq 1,$$
or
$$\liminf_{n\to\infty} 2^{-n} A_n / B_n < 1.$$

In the first case we can choose k so large that $B_i \leq 2 \cdot 2^{-i} A_i$ for $i \geq k$. It follows that for $i \geq k$
$$\frac{a_i 2^{is}}{2^{-i} A_i + B_i} \geq \frac{a_i 2^{is}}{3 \cdot 2^{-i} A_i} = \frac{a_i 2^{i(s+1)}}{3 A_i},$$
and thus
$$\liminf_{n\to\infty} \sum_{i=k}^{n} \frac{a_i 2^{is}}{2^{-i} A_i + B_i} \geq \liminf_{n\to\infty} \sum_{i=k}^{n} \frac{a_i 2^{i(s+1)}}{3 A_i}$$
$$\geq \liminf_{n\to\infty} \frac{1}{3 A_n} \sum_{i=k}^{n} a_i 2^{i(s+1)} = \liminf_{n\to\infty} \frac{A_n - A_{k-1}}{3 A_n} = \tfrac{1}{3}.$$

In the second case there are arbitrarily large k such that $2^{-k} A_k / B_k \leq 2$, and then for $i \geq k$ we have
$$2^{-i} A_i + B_i = 2^{-i} A_k + \sum_{j=k+1}^{i} 2^{j-i} a_j 2^{js} + B_i$$
$$\leq 2^{-k} A_k + \sum_{j=k+1}^{i} a_j 2^{js} + B_i = 2^{-k} A_k + B_k \leq 3 B_k.$$

It follows that
$$\liminf_{n\to\infty} \sum_{i=k}^{n} \frac{a_i 2^{is}}{2^{-i} A_i + B_i} \geq \liminf_{n\to\infty} \frac{1}{3 B_k} \sum_{i=k}^{n} a_i 2^{is} \geq \liminf_{n\to\infty} \frac{B_k - B_n}{3 B_k} = \tfrac{1}{3},$$
which proves the lemma. $\square$

Now let $f \in YL(E)$, $f = \sum_{i=0}^{\infty} f_i$, as in Definition 1.1.5, and denote the partial sums by

$$\Phi_i = \sum_{j=0}^{i} f_j. \tag{2.2.3}$$

Set

$$d_{i,k} = \max\{ 2^{-i|\beta|}|D^\beta \Phi_i(x)| : x \in Q_{i,k}(3), |\beta| \leq M \}, \tag{2.2.4}$$

where $M > \varepsilon_-$.

LEMMA 2.2.3. *Let the functions f, $\{f_i\}_{i=0}^{\infty}$, $\{g_i\}_{i=0}^{\infty}$, $\{h_i\}_{i=0}^{\infty}$, and the set A satisfy the conditions of Corollary 2.1.6. Then*

$$\sum_{i=0}^{\infty} \sum_{k \in \mathbf{Z}^N} \frac{t_{i,k}}{d_{i,k}} \chi_{i,k}(x) = \infty. \tag{2.2.5}$$

for all $x \in A$.

PROOF. Set

$$F_s = \{ x : \sum_{i=0}^{\infty} 2^{is} h_i(x) = \infty \}, \quad s = 0, 1, \ldots.$$

Then

$$A = (A \cap F_0) \bigcup \left(\cup_{s=0}^{[\varepsilon_-]} A \cap (F_{s+1} \setminus F_s) \right),$$

since $F_s = \mathbf{R}^N$ for $s > \varepsilon_-$ by condition (v) in Lemma 2.1.5.

We first observe that (2.2.5) holds for all $x \in F_0$. In fact, for all (i,k)

$$d_{i,k} = \max_{y \in Q_{i,k}(3), |\beta| \leq M} 2^{-i|\beta|} |D^\beta \Phi_i(y)|$$

$$\leq \sum_{j=0}^{i} \max_{y \in Q_{i,k}(3), |\beta| \leq M} 2^{-i|\beta|} |D^\beta f_j(y)| \leq \max_{|\beta| \leq M} \sum_{Q_{j,l} \supset Q_{i,k}} 2^{(j-i)|\beta|} s_{j,l},$$

and thus

$$d_{i,k} \leq \sum_{Q_{j,l} \supset Q_{i,k}} s_{j,l} = \sum_{j=0}^{i} g_i(x) \leq \sum_{j=0}^{i} h_i(x)$$

for $x \in Q_{i,k}$. The claim follows directly from Lemma 2.2.2(a).

Now let $x \in A \cap (F_{s+1} \setminus F_s)$, $0 \leq s \leq \varepsilon_-$, so that $\sum_{i=0}^{\infty} 2^{is} h_i(x) < \infty$, and $\sum_{i=0}^{\infty} 2^{i(s+1)} h_i(x) = \infty$.

The key assumption is that $P_x^s f = 0$, i.e.,

$$\sum_{i=0}^{\infty} D^\beta f_i(x) = 0 \quad \text{for all } \beta, |\beta| \leq s. \tag{2.2.6}$$

We claim that under this assumption, if $x \in Q_{i,k}$, then

$$2^{is} d_{i,k} \leq C 2^{-i} \sum_{j=0}^{i} 2^{j(s+1)} g_j(x) + C \sum_{j=i+1}^{\infty} 2^{js} g_j(x). \tag{2.2.7}$$

Then the result follows from (iii) in Lemma 2.2.2, since $g_j \leq h_j$.

First let β be a multi-index with $s+1 \leq |\beta| \leq M$. Then, for any $y \in Q_{i,k}(3)$, we have

$$(2.2.8) \qquad 2^{-i|\beta|}|D^\beta \Phi_i(y)| \leq \sum_{j=0}^{i} 2^{(j-i)|\beta|} g_j(y) \leq \sum_{j=0}^{i} 2^{(j-i)(s+1)} g_j(y).$$

Now let $|\beta| \leq s$. Then, the crucial observation is that by (2.2.6)

$$(2.2.9) \qquad |D^\beta \Phi_i(x)| = \left| \sum_{j=0}^{i} D^\beta f_j(x) \right| = \left| \sum_{j=i+1}^{\infty} D^\beta f_j(x) \right| \leq \sum_{j=i+1}^{\infty} 2^{j|\beta|} g_j(x).$$

As we already noted, this is the only place in the entire proof of Theorem 2.2.1 where the assumption (2.2.2) is used.

Expanding $D^\beta \Phi_i(y)$ by Taylor's formula at $x \in Q_{i,k}$ we obtain for $y \in Q_{i,k}(3)$,

$$2^{-i|\beta|}|D^\beta \Phi_i(y)|$$
$$\leq C \sum_{|\beta| \leq |\gamma| \leq s} 2^{-i|\gamma|} |D^\gamma \Phi_i(x)| + C \sum_{|\gamma|=s+1} \max_{0 \leq t \leq 1} 2^{-i|\gamma|} |D^\gamma \Phi_i(x + t(y-x))|.$$

and by (2.2.9) and (2.2.8)

$$2^{-i|\beta|}|D^\beta \Phi_i(y)|$$
$$\leq C \sum_{|\beta| \leq |\gamma| \leq s} 2^{-i|\gamma|} \sum_{j=i+1}^{\infty} 2^{j|\gamma|} g_j(x) + C \sum_{j=0}^{i} 2^{(j-i)(s+1)} g_j(x)$$
$$\leq C \sum_{j=i+1}^{\infty} 2^{(j-i)s} g_j(x) + C \sum_{j=0}^{i} 2^{(j-i)(s+1)} g_j(x).$$

The inequality (2.2.7) follows from this and (2.2.8), and Lemma 2.2.3 is proved. $\square$

PROOF OF THEOREM 2.2.1. We only have to prove that (i) in the theorem implies (iii). Let f, $\{f_i\}$, $\{g_i\}$, $\{h_i\}$ be functions satisfying the conditions in Corollary 2.1.6, and let $\{d_{i,k}\}$ be the numbers defined by (2.2.4). We use (2.2.5) to construct a multiplier η.

Recall that $h_i = \sum_{k \in \mathbf{Z}^N} t_{i,k} \chi_{i,k}$. Let $\varphi \geq 0$ be a function in $C_0^\infty(\mathbf{R}^N)$ such that $\operatorname{supp} \varphi \subset Q_{0,0}(2)$ and $\varphi(x) \geq 1$ on $Q_{0,0}(\frac{3}{2})$. Define $\varphi_{i,k}(x) = \varphi(2^i x - k)$ for $i \in \mathbf{N}_0$ and $k \in \mathbf{Z}^N$, and set

$$\kappa_i(x) = \sum_{k \in \mathbf{Z}^N} \min\left\{1, \frac{t_{i,k}}{d_{i,k}}\right\} \varphi_{i,k}(x).$$

Then fix a large number i_0, and denote

$$K_i = \sum_{j=i_0}^{i} \kappa_j, \quad i \geq i_0.$$

Let $\Psi \in C^\infty(\mathbf{R})$ be a function such that $\Psi(0) = 0$, $0 \leq \Psi(t) \leq 1$, $\Psi(t) = 1$ for all $t \geq 1$, and set

$$\eta = \lim_{i \to \infty} \Psi(K_i) = \Psi\left(\sum_{i=i_0}^{\infty} \kappa_i \right),$$

Then Lemma 2.2.3 implies that $\eta(x) = 1$ in a neighborhood of A. In fact, $\sum_{i=i_0}^{\infty} \kappa_i$ is lower semicontinuous.

We shall finish the proof of Theorem 2.2.1 by showing that we can make $\|\eta f\|_{YL(E)}$ arbitrarily small by choosing i_0 sufficiently large. By Theorem 1.1.14, assumption (2.2.1), and Lemma 1.1.3 it is enough to prove that there are functions u_i, $i \geq i_0$, and a constant C, such that

$$\eta f = \sum_{i=i_0}^{\infty} u_i,$$

and such that for for all $|\beta| \leq M$

(2.2.10) $$|D^\beta u_i| \leq 2^{i|\beta|} \sum_{k \in \mathbf{Z}^N} t_{i,k} \widetilde{\chi}_{i,k},$$

where $\widetilde{\chi}_{i,k} = \chi(Q_{i,k}(3))$.

Setting
$$\eta_{i_0} = \Psi(\kappa_{i_0}),$$
$$\eta_i = \Psi(K_i) - \Psi(K_{i-1}), \quad i \geq i_0 + 1,$$

we have $\Psi(K_i) = \sum_{j=i_0}^{i} \eta_j$, and $\eta = \sum_{j=i_0}^{\infty} \eta_j$.

We have

$$\eta f = \sum_{i=i_0}^{\infty} \eta_i \sum_{j=0}^{\infty} f_j = \sum_{i=i_0}^{\infty} \eta_i \left(\sum_{j=0}^{i} f_j + \sum_{j=i+1}^{\infty} f_j \right)$$
$$= \sum_{i=i_0}^{\infty} \eta_i \Phi_i + \sum_{j=i_0+1}^{\infty} f_j \sum_{i=i_0}^{j-1} \eta_i = \sum_{i=i_0}^{\infty} \eta_i \Phi_i + \sum_{i=i_0+1}^{\infty} f_i \Psi(K_{i-1}).$$

We set
$$u_i = \eta_i \Phi_i + \Psi(K_{i-1}) f_i,$$

and we shall estimate $D^\beta u_i$ for $|\beta| \leq M$. We first observe that

(2.2.11) $$|D^\beta \kappa_i(x)| \leq C\, 2^{i|\beta|} \sum_{k \in \mathbf{Z}^N} \min\left\{1, \frac{t_{i,k}}{d_{i,k}}\right\} \widetilde{\chi}_{i,k}(x),$$

and that

(2.2.12) $$|D^\beta K_i(x)| \leq \sum_{j=i_0}^{i} |D^\beta \kappa_j(x)| \leq C \sum_{j=i_0}^{i} 2^{j|\beta|} \leq C\, 2^{i|\beta|}.$$

We write

(2.2.13) $$\eta_i = \Psi(K_i) - \Psi(K_{i-1}) = \kappa_i \int_0^1 \Psi'(K_{i-1} + \kappa_i t)\, dt = \kappa_i Y_i.$$

By repeatedly using the chain rule, (2.2.11), and (2.2.12), we find

$$|D^\beta Y_i(x)| \leq C\, 2^{i|\beta|}.$$

It follows from (2.2.13), using the Leibniz rule and (2.2.11), that

$$|D^\beta \eta_i(x)| \leq C 2^{i|\beta|} \sum_{k \in \mathbf{Z}^N} \frac{t_{i,k}}{d_{i,k}} \widetilde{\chi}_{i,k}(x).$$

For an $x \in Q_{i,l}$ the last sum contains only those 3^N terms $(t_{i,k}/d_{i,k})\widetilde{\chi}_{i,k}(x)$ for which $Q_{i,k}(3)$ intersects $Q_{i,l}$. By (2.2.4)

$$|D^\beta \Phi_i(x)| \leq 2^{i|\beta|} \min_{Q_{i,k}(3) \cap Q_{i,l} \neq \emptyset} d_{i,k},$$

and thus, again by the Leibniz rule,

$$|D^\beta(\eta_i \Phi_i)| \leq C\, 2^{i|\beta|} \sum_{k \in \mathbf{Z}^N} t_{i,k} \widetilde{\chi}_{i,k}$$

for $|\beta| \leq M$.

Finally, by the chain rule and (2.2.12)

$$|D^\beta \Psi(K_{i-1})| \leq C\, 2^{i|\beta|},$$

and by (2.1.3)

$$|D^\beta f_i| \leq 2^{i|\beta|} \sum_{k \in \mathbf{Z}^N} s_{i,k} \chi_{i,k},$$

so again by the Leibniz rule

$$|D^\beta(\Psi(K_{i-1}) f_i)| \leq 2^{i|\beta|} \sum_{k \in \mathbf{Z}^N} s_{i,k} \chi_{i,k}.$$

This proves the claim (2.2.10) and finishes the proof of Theorem 2.2.1. □

2.3. Spectral Synthesis in Spaces of Distributions

In order to formulate the spectral synthesis problem in the distribution spaces $Y(E)$ we have to recall a few facts about $Y(E)$ and $YL(E)$. We assume that $E \in S(\varepsilon_+, \varepsilon_-, r)$, where $0 < r < 1$ and $0 < \varepsilon_+ \leq N/r - N$. Note that $YL(E) = Y(E)$ if $\varepsilon_+ > N \max\{\frac{1}{r} - 1, 0\}$ by Proposition 1.1.12.

Let $f \in Y(E)$. If Ψ is defined as in (1.3.2) with $\Psi_t(x) = t^{-N}\Psi(x/t)$ for $t > 0$, we know that $f = \lim_{t \to 0} f * \Psi_t$ in the sense of distributions. Moreover, if we set $\psi_0 = \Psi$, $\psi_j = \Psi_{2^{-j}} - \Psi_{2^{-j+1}}$ for $j \geq 1$, and $f_j = f * \psi_j$, it follows that

$$f = \lim_{j \to \infty} f * \Psi_{2^{-j}} = \sum_{j=0}^{\infty} f_j,$$

and by Theorem 1.1.15 we have $\|\{f_i\}_0^\infty\|_E \leq C \|f\|_{Y(E)}$.

By Theorem 1.1.14 we also know that there is a function $\tilde{f} \in YL(E)$ such that $\lim_{j \to \infty} f * \Psi_{2^{-j}}(x) = \sum_{j=0}^{\infty} f_j(x) = \tilde{f}(x)$ with convergence in $L_{r,\text{loc}}$, and pointwise convergence at all L_r-Lebesgue points of $\tilde{f}$. Moreover, $\tilde{f}$ has derivatives $D^\beta \tilde{f}(x)$ for $|\beta| \leq \varepsilon_-$ and L_r-differentials in the sense of Theorem 2.1.4.

We extend this result to convolutions with more general functions Φ.

LEMMA 2.3.1. *Let E, f, and $\tilde{f}$ be as above. Let $\Phi \in \mathcal{S}$, $\Phi_t = t^{-N}\Phi(\,\cdot\,/t)$, and suppose that $\int_{\mathbf{R}^N} \Phi\, dx = 1$. For every integer s, $0 \leq s \leq \varepsilon_-$, there is a set F_s independent of Φ, such that $\operatorname{cap}(F_s, s, E) = 0$, and $\lim_{t \to 0} D^\beta f * \Phi_t(x) = D^\beta \tilde{f}(x)$ for all $|\beta| = s$ and $x \in F_s^c$.*

PROOF. The proof is the same for all β, so it is enough to set $\beta = 0$. It is clearly sufficient to consider $\lim_{i \to \infty} f * \Phi_{2^{-i}}$. Set

$$A_K(\Phi) = \max_{|\alpha| \leq K} \max_x (1 + |x|)^K |D^\alpha \Phi(x)|$$

for some large $K > 0$. Let $\{\psi_j\}_{j=0}^\infty$ be the functions defined above, and let $\{h_j\}_{j=0}^\infty$ be the functions defined in Theorem 1.1.15(ii) for a λ satisfying the conditions of that theorem.

We claim that if $K > \lambda$ is sufficiently large, then
$$|(f * \Phi_{2^{-i}}) * \psi_j(x)| = |f * (\Phi_{2^{-i}} * \psi_j)(x)| \leq CA_K(\Phi)h_j(x) \quad \text{for all } i \geq 0.$$

We observe that $\operatorname{supp} \mathcal{F}\psi_j \subset B(0, 2^{j+1}) \setminus B(0, 2^{j-1})$. It follows, in particular, that the functions $\Phi_{2^{-i}} * \psi_j$ have zero moments of all orders for $j \geq 1$. By Lemma 1.2.9
$$|D^\alpha(\Phi_{2^{-i}} * \psi_j)(x)| \leq (\Phi_{2^{-i}} * |D^\alpha \psi_j|)(x) \leq \frac{CA_K(\Phi)2^{j(N+|\alpha|)}}{(1+2^j|x|)^\lambda} \quad \text{for } i \geq j,$$
which implies the claim in that case.

In order to get a suitable estimate for large j we notice that the assumption on Φ implies that for any sufficiently large K there is a C so that for all $\xi \in \operatorname{supp} \mathcal{F}\psi_j$
$$\max_{|\beta| \leq \lambda} (1+|\xi|)^\lambda |D^\beta \mathcal{F}\Phi(\xi)| \leq CA_K(\Phi)2^{-j(\lambda+N)} \quad \text{for all } j \geq 1.$$

It is easily seen that there is a function $\Phi^{(j)} \in \mathcal{S}$ such that $\mathcal{F}\Phi^{(j)}(\xi) = \mathcal{F}\Phi(\xi)$ on $\operatorname{supp} \mathcal{F}\psi_j$, whence $\Phi * \psi_j = \Phi^{(j)} * \psi_j$, and such that
$$\max_{|\beta| \leq \lambda} \max_{\xi \in \mathbf{R}^N} (1+|\xi|)^\lambda |D^\beta \mathcal{F}\Phi^{(j)}(\xi)| \leq CA_K(\Phi)2^{-j(\lambda+N)} \quad \text{for all } j \geq 1.$$

It follows readily that $A_\lambda(\Phi^{(j)}) \leq CA_K(\Phi)2^{-j\lambda}$, and hence by Lemma 1.2.9
$$|\Phi * D^\alpha \psi_j(x)| \leq \frac{CA_K(\Phi)2^{-j\lambda}2^{j|\alpha|}}{(1+|x|)^\lambda} \quad \text{for } j \geq 1.$$

By scaling,
$$|\Phi_{2^{-i}} * D^\alpha \psi_j(x)| \leq \frac{CA_K(\Phi)2^{-j\lambda}2^{j|\alpha|}2^{iN}}{(1+2^i|x|)^\lambda} \quad \text{for } i \leq j.$$

But
$$\frac{2^{-j\lambda}2^{iN}}{(1+2^i|x|)^\lambda} \leq 2^{-i(\lambda-N)}\min\{1,(2^j|x|)^{-\lambda}\} \leq \frac{C}{(1+2^j|x|)^\lambda}$$
for $i \leq j$ and $\lambda > N$, which proves the claim.

The functions h_j are independent of Φ, and by Theorem 1.1.15(ii)
$$\|\{h_j\}_{j=0}^\infty\|_E \leq CA_K(\Phi)\|f\|_{Y(E)}.$$

By Lemma 2.1.2 there is a set F with $\operatorname{cap}(F,0,E) = 0$ such that $\sum_{j=0}^\infty h_j(x) < \infty$ for all $x \in F^c$. Since $(f * \Phi_{2^{-i}}) * \psi_j = f_j * \Phi_{2^{-i}}$, and $\lim_{i \to \infty} f_j * \Phi_{2^{-i}}(x) = f_j(x)$ for all x, it follows easily that
$$\lim_{i \to \infty} f * \Phi_{2^{-i}}(x) = \sum_{j=0}^\infty f_j(x) = \tilde{f}(x) \quad \text{for all } x \in F^c,$$
which proves the lemma. $\square$

It is not known whether Theorem 2.2.1 can be extended to the general distribution spaces $Y(E)$, when these are different from the corresponding $YL(E)$.

We have the following theorem, whose proof depends on properties of the norm in B-spaces. Recall that the space $B_{p,\theta}^l$ is identified with $Y(E)$ for $E = l_\theta^l(L_p)$, so that $E \in S(l,l,r)$ for any $r \in (0,p)$. See the examples following Definition 1.1.1.

2.3. SPECTRAL SYNTHESIS IN SPACES OF DISTRIBUTIONS

By Definition 2.1.1 we associate s-capacities associated to the space E, denoted by $\mathrm{cap}(\,\cdot\,, s, E)$. It is easily seen that for $s \in \mathbf{N}$ the capacity $\mathrm{cap}(\,\cdot\,, s, l_\theta^l(L_p))$ is equivalent to $\mathrm{cap}(\,\cdot\,, 0, l_\theta^{l-s}(L_p))$. We write $\mathrm{cap}(\,\cdot\,, B_{p,\theta}^l)$ for $\mathrm{cap}(\,\cdot\,, 0, l_\theta^l(L_p))$.

THEOREM 2.3.2. *Let* $A \subset \mathbf{R}^N$, $0 < p < 1$, $0 < \theta < \infty$, $0 < l < \frac{N}{p} - N$, *and* $\frac{N}{p} - l \notin \mathbf{Z}$. *Let* $f \in B_{p,\theta}^l$, *and suppose that the following conditions are satisfied:*
  (i) $\mathrm{supp}\, f \subset \mathbf{R}^N \setminus \mathrm{int}\, A$;
  (ii) *For some* $\Phi \in \mathcal{S}$ *such that* $\int \Phi \, dx \neq 0$, *and all multi-indices* β, $|\beta| \leq l$,
$$\lim_{t \to 0} D^\beta f * \Phi_t(x) = 0$$
for q.e. $x \in \partial A$ *with respect to the capacity* $\mathrm{cap}(\,\cdot\,, B_{p,\theta}^{i-|\beta|})$.

Then, for all $\varepsilon > 0$ *there exists* $f_\varepsilon \in B_{p,\theta}^l$ *such that* $\mathrm{supp}\, f_\varepsilon \cap A = \emptyset$, *and* $\|f - f_\varepsilon\|_{B_{p,\theta}^l} < \varepsilon$.

REMARK. It is unknown whether Theorem 2.2.1(iii) can be extended to this situation.

We first prove the following theorem.

THEOREM 2.3.3. *Let* $0 < p < 1$, $0 < \theta \leq \infty$, $0 < l < \frac{N}{p} - N$ *and suppose that* $\frac{N}{p} - l \notin \mathbf{Z}$. *Let* $u \in BL_{p,\theta}^l$. *Then there is a distribution* $v \in B_{p,\theta}^l$, *and a constant* C *independent of* u, *such that:*
  (i) $\widetilde{D^\beta v} = D^\beta u$ *for all multi-indices* β, $|\beta| \leq l$, *i.e.,*
$$\lim_{t \to 0} D^\beta v * \Phi_t(x) = D^\beta u(x)$$
q.e. with respect to the capacity $\mathrm{cap}(\,\cdot\,, B_{p,\theta}^{l-|\beta|})$ *for all* $\Phi \in \mathcal{S}$ *such that* $\int_{\mathbf{R}^N} \Phi \, dx = 1$;
  (ii) *if there is an open set* Ω *and a* $g \in B_{p,\theta}^l$ *such that* $u = \tilde{g}$ *in* Ω, *then* $v = g$ *in* Ω *in the sense that* $\varphi v = \varphi g$ *for all* $\varphi \in C_0^\infty(\Omega)$, *in particular* $\mathrm{supp}\, v \subset \mathrm{supp}\, u$;
  (iii) $\|v\|_{B_{p,\theta}^l} \leq C \|u\|_{BL_{p,\theta}^l}$.

PROOF. We set $L = [\frac{N}{p} - N - l] + 1$, where $[\,\cdot\,]$ denotes the integral part, and observe that by the assumptions
$$0 \leq L - 1 < \frac{N}{p} - N - l < L.$$

We also fix a number r so that

(2.3.1) $\qquad 0 < r < p \quad \text{and} \quad \frac{N}{r} - N - l < L.$

By Theorem 1.1.15 we can assume that u has a representation
$$u = \sum_{i=0}^\infty u_i, \quad \text{where} \quad \|\{2^{il} u_i\}_0^\infty\|_{l_\theta(L_p)} \leq 2 \|u\|_{BL_{p,\theta}^l},$$
$$u_i = \sum_{k \in \mathbf{Z}^N} s_{i,k} a_{i,k}, \quad \text{and} \quad a_{i,k} \in C_0^M(Q_{i,k}(3)),$$

for some $M > l$, as in (2.1.1)–(2.1.4). Moreover, the constants $s_{i,k}$ are such that the functions $g_i = \sum_{k \in \mathbf{Z}^N} |s_{i,k}| \chi_{i,k}$ satisfy
$$\|\{2^{il} g_i\}_0^\infty\|_{l_\theta(L_p)} \leq C \|u\|_{BL_{p,\theta}^l}.$$

For each $Q_{i,k}$ such that $Q_{i,k}(3)$ intersects $\operatorname{supp} u$ we choose a point $x_{i,k} \in Q_{i,k}(3) \cap \operatorname{supp} u$. For all multi-indices β with $|\beta| \leq L-1$ we set

$$b_{\beta;i,k} = \frac{(-1)^{|\beta|}}{\beta!} \int_{\mathbf{R}^N} (x-x_{i,k})^\beta a_{i,k}(x)\, dx,$$

and we define distributions $c_{i,k}$ by setting

$$c_{i,k} = a_{i,k} - \sum_{|\beta| \leq L-1} b_{\beta;i,k} D^{(\beta)} \delta(x_{i,k}),$$

where $D^{(\beta)}\delta(x_{i,k})$ denotes derivatives of the Dirac measure at $x_{i,k}$. Then clearly $c_{i,k} \perp \mathfrak{P}_{L-1}$, and also

(2.3.2) $$|b_{\beta;i,k}| \leq C 2^{-i(|\beta|+N)}.$$

We note that $c_{i,k} = a_{i,k}$ if $a_{i,k} \perp \mathfrak{P}_{L-1}$. In particular, if the assumption in (ii) is satisfied for some open Ω, we can assume that $c_{i,k} = a_{i,k}$ for all (i,k) such that $Q_{i,k}(3) \subset \Omega$. We define

(2.3.3) $$v_i = \sum_{k \in \mathbf{Z}^N} s_{i,k} c_{i,k},$$

and we claim that $\sum_{i=0}^\infty v_i$ converges to a distribvtion $v \in B_{p,\theta}^l$ which satisfies the reqvirements of the theorem.

We let $\psi \in C_0^\infty(B(0,1))$ satisfy $\|\psi\|_{C^L} \leq 1$, and $\psi \perp \mathfrak{P}_{M-1}$, and set $\psi_j(x) = 2^{jN}\psi(2^j x)$. We denote

(2.3.4) $$V_j(x) = \sum_{i=0}^\infty \sup_\psi |v_i * \psi_j(x)|, \qquad j \in \mathbf{N}_0.$$

By (ii') in Theorem 1.1.15 the above claim follows if we prove that

(2.3.5) $$\left\|\{2^{jl} V_j\}_{j=0}^\infty\right\|_{l_\theta(L_p)} < C \left\|\{2^{jl} g_j\}_{j=0}^\infty\right\|_{l_\theta(L_p)}.$$

We proceed to prove this inequality, much as in the proof of the implication (iii) $\Rightarrow$ (ii) in Theorem 1.1.15.

We set

(2.3.6) $$V_{i,j}(x) = \sup_\psi \left| \sum_{k \in \mathbf{Z}^N} s_{i,k} c_{i,k} * \psi_j(x) \right|,$$

and we first consider the case $i \leq j$. We have

(2.3.7) $$V_{i,j}(x) \leq \sup_\psi \left| \sum_{k \in \mathbf{Z}^N} s_{i,k} a_{i,k} * \psi_j(x) \right|$$
$$+ \sup_\psi \sum_{k \in \mathbf{Z}^N} \left| s_{i,k} \sum_{|\beta| \leq L-1} b_{\beta;i,k} D^{(\beta)} \delta(x_{i,k}) * \psi_j(x) \right|.$$

Since $\operatorname{supp} a_{i,k} * \psi_j \subset Q_{i,k}(5)$ it follows from Lemma 1.2.9 that

$$|a_{i,k} * \psi_j(x)| \leq C 2^{-(j-i)M} \chi(Q_{i,k}(5))(x),$$

and consequently

$$\sup_\psi \left| \sum_{k \in \mathbf{Z}^N} s_{i,k} a_{i,k} * \psi_j(x) \right| \leq C 2^{-(j-i)M} \sum_{k \in \mathbf{Z}^N} |s_{i,k}| \chi(Q_{i,k}(5))(x).$$

The contribution of this last quantity to the norm can be estimated by means of (1.1.6), but in the present case we also have the elementary estimate,

$$(2.3.8) \quad \int_{\mathbf{R}^N} \left(\sum_{k \in \mathbf{Z}^N} |s_{i,k}| \chi(Q_{i,k}(5))(x)\right)^p dx \leq C 2^{-iN} \sum_{k \in \mathbf{Z}^N} |s_{i,k}|^p$$

$$= C \int_{\mathbf{R}^N} \left(\sum_{k \in \mathbf{Z}^N} |s_{i,k}|^p \chi_{i,k}\right) dx = C \int_{\mathbf{R}^N} g_i^p \, dx,$$

Further, by (2.3.2),

$$\sum_{|\beta| \leq L-1} |b_{\beta;i,k} D^{(\beta)} \delta(x_{i,k}) * \psi_j(x)| = \sum_{|\beta| \leq L-1} |b_{\beta;i,k}| |D^{(\beta)} \psi_j(x - x_{i,k})|$$

$$\leq C \chi(B(x_{i,k}, 2^{-j}))(x) \sum_{|\beta| \leq L-1} 2^{(j-i)(|\beta|+N)}$$

$$\leq C 2^{(j-i)(N+L-1)} \chi(B(x_{i,k}, 2^{-j}))(x),$$

so that in (2.3.7)

$$(2.3.9) \quad \sup_\psi \sum_{k \in \mathbf{Z}^N} \left| s_{i,k} \sum_{|\beta| \leq L-1} b_{\beta;i,k} D^{(\beta)} \delta(x_{i,k}) * \psi_j(x) \right|$$

$$\leq C 2^{(j-i)(N+L-1)} \sum_{k \in \mathbf{Z}^N} |s_{i,k}| \chi(B(x_{i,k}, 2^{-j}))(x).$$

It is in order to estimate the norm of this last quantity that we need the special properties of the $l_\theta(L_p)$-norm.

We find

$$(2.3.10) \quad \int_{\mathbf{R}^N} \left(\sum_{k \in \mathbf{Z}^N} |s_{i,k}| \chi(B(x_{i,k}, 2^{-j}))(x)\right)^p dx \leq C 2^{-jN} \sum_{k \in \mathbf{Z}^N} |s_{i,k}|^p$$

$$= C 2^{(i-j)N} \int_{\mathbf{R}^N} \left(\sum_{k \in \mathbf{Z}^N} |s_{i,k}|^p \chi_{i,k}\right) dx = C 2^{(i-j)N} \int_{\mathbf{R}^N} g_i^p \, dx,$$

where the first inequality is an immediate consequence of the fact that $j \geq i$. Thus, combining (2.3.10) with (2.3.8) in (2.3.7) we obtain

$$(2.3.11) \quad \|V_{i,j}\|_{L_p} \leq C\left(2^{-(j-i)M} + 2^{(j-i)(N-\frac{N}{p}+L-1)}\right) \|g_i\|_{L_p}, \quad j \geq i.$$

Turning to the estimate of $V_{i,j}$ for $i > j$ we fix an x and estimate

$$\left(a_{i,k} - \sum_{|\beta| \leq L-1} b_{\beta;i,k} D^{(\beta)} \delta(x_{i,k})\right) * \psi_j(x) = c_{i,k} * \psi_j(x) = \langle c_{i,k}, \psi_j(x - \cdot) \rangle$$

for those k for which $\mathrm{supp}\, c_{i,k}$ intersects $B(x, 2^{-j})$. It is no loss of generality to assume that $x = 0$ and estimate $\langle c_{i,k}, \psi_j \rangle$. We denote by $\pi_{i,k}$ the Taylor polynomial of degree $L-1$ of ψ_j at $x_{i,k}$, and notice that for $y \in \mathrm{supp}\, c_{i,k}$

$$|\psi_j(y) - \pi_{i,k}(y)| \leq C|y - x_{i,k}|^L \max_\xi \sum_{|\gamma|=L} |D^\gamma \psi_j(\xi)| \leq C 2^{-iL} 2^{j(N+L)}.$$

By assumption

$$\langle c_{i,k}, \psi_j \rangle = \langle c_{i,k}, \psi_j - \pi_{i,k} \rangle$$

$$= \langle a_{i,k}, \psi_j - \pi_{i,k} \rangle - \sum_{|\beta| \leq L-1} b_{\beta;i,k} \langle D^{(\beta)} \delta(x_{i,k}), \psi_j - \pi_{i,k} \rangle.$$

But here the last sum vanishes, since
$$\langle D^{(\beta)}\delta(x_{i,k}), \pi_{i,k}\rangle = (-1)^{|\beta|} D^\beta \psi(x_{i,k}) = \langle D^{(\beta)}\delta(x_{i,k}), \psi_j\rangle.$$
Consequently
$$|\langle c_{i,k}, \psi_j\rangle| = |\langle a_{i,k}, \psi_j - \pi_{i,k}\rangle| \leq C 2^{-iL} 2^{j(N+L)} \int_{\mathbf{R}^N} |a_{i,k}|\, dy \leq C 2^{-(i-j)(N+L)}.$$
Allowing x to be arbitrary, it follows that
$$V_{i,j}(x) \leq C 2^{-(i-j)(N+L)} \sum_{\{k: Q_{i,k}(3) \cap B(x, 2^{-j}) \neq \emptyset\}} |s_{i,k}|.$$
Here the last sum is estimated by
$$\Big(\sum_k |s_{i,k}|\Big)^r \leq \sum_k |s_{i,k}|^r \leq 2^{iN} \int_{B(x, 2^{-j+1}\sqrt{N})} g_i^r\, dy \leq C 2^{(i-j)N} M_r g_i(x)^r,$$
with r given by (2.3.1), and thus
(2.3.12) $$V_{i,j}(x) \leq C 2^{-(i-j)(N-\frac{N}{r}+L)} M_r g_i(x).$$
We now choose $\iota > 0$ so that
$$\iota \leq N - \frac{N}{r} + L + l,$$
$$\iota \leq 1 + \frac{N}{p} - N - L - l, \quad \text{and}$$
$$\iota \leq M - l.$$
It follows from (2.3.11), (2.3.12), and the Hardy–Littlewood–Wiener theorem that
$$2^{jl}\|V_{i,j}\|_{L_p} \leq C 2^{-(i-j)\iota} 2^{il} \|g_i\|_{L_p}, \quad j \leq i, \quad \text{and}$$
$$2^{jl}\|V_{i,j}\|_{L_p} \leq C 2^{-(j-i)\iota} 2^{il} \|g_i\|_{L_p}, \quad j \geq i.$$
Setting $g_i = 0$ for $i < 0$, we obtain
$$\big\|\{2^{jl} V_{j-s,j}\}_{j=s}^\infty\big\|_{l_\theta(L_p)} \leq C 2^{-s\iota} \big\|\{2^{il} g_i\}_{i=0}^\infty\big\|_{l_\theta(L_p)}, \quad \text{and}$$
$$\big\|\{2^{jl} V_{j+s,j}\}_{j=0}^\infty\big\|_{l_\theta(L_p)} \leq C 2^{-s\iota} \big\|\{2^{il} g_i\}_{i=0}^\infty\big\|_{l_\theta(L_p)}.$$
But
$$\{2^{jl} V_j\}_{j=0}^\infty = \sum_{s=0}^\infty \{2^{jl} V_{j-s,j}\}_{j=s}^\infty + \sum_{s=1}^\infty \{2^{jl} V_{j+s,j}\}_{j=l}^\infty,$$
and it follows that
$$\big\|\{2^{jl} V_j\}_{j=0}^\infty\big\|_{l_\theta(L_p)}^\kappa \leq C \big\|\{2^{jl} g_j\}_{j=0}^\infty\big\|_{l_\theta(L_p)}^\kappa \sum_{s=0}^\infty 2^{-s\iota\kappa} = C\big\|\{2^{jl} g_j\}_{j=0}^\infty\big\|_{l_\theta(L_p)}^\kappa,$$
where $\kappa = \min\{p, \theta\}$. By Theorem 1.1.15 this proves the claim that $v = \sum_{i=0}^\infty v_i$ belongs to $B_{p,\theta}^l$ and satisfies $\|v\|_{B_{p,\theta}^l} \leq C\|u\|_{BL_{p,\theta}^l}$.

It is obvious from the construction that the claim (ii) is satisfied, in particular that $\operatorname{supp} v \subset \operatorname{supp} u$.

We finally prove (i). (Note that this property is not needed for the proof of Theorem 2.3.2.) It is enough to assume $\beta = 0$.

It is clear that (i) holds for each v_i, in fact, $\lim_{t \to 0} v_i * \Phi_t(x) = u_i(x)$ for all x not coinciding with $x_{i,k}$ for some k.

2.3. SPECTRAL SYNTHESIS IN SPACES OF DISTRIBUTIONS

Let $n = 1, 2, \ldots$, and choose for each n an integer i_n so large that
$$\left\|\{2^{il}g_i\}_{i=i_n}^\infty\right\|_{l_\theta(L_p)} < 4^{-n}.$$
This is possible since $\theta < \infty$. It follows from (2.3.5) with $g_i = 0$ for $0 \le i < i_n$ that $\sum_{i=i_n}^\infty v_i = V^{(n)}$ satisfies $\|V^{(n)}\|_{B^l_{p,\theta}} < C4^{-n}$.

We observe that by Lemma 2.3.1 there is a function $\widetilde{V^{(n)}} \in BL^l_{p,\theta}$ with
$$\|\widetilde{V^{(n)}}\|_{BL^l_{p,\theta}} < C\|V^{(n)}\|_{B^l_{p,\theta}} < C4^{-n},$$
such that
$$\lim_{t\to 0} V^{(n)} * \Phi_t(x) = \widetilde{V^{(n)}}(x)$$
q.e. in the sense of $\mathrm{cap}(\,\cdot\,, B^l_{p,\theta})$. Moreover, by the definition of capacity there is a set F_n such that the sets where $|\widetilde{V^{(n)}}(x)| > 2^{-n}$ or $\sum_{i=i_n}^\infty |u_i(x)| > 2^{-n}$ are contained in F_n, and $\mathrm{cap}\, F_n < C2^{-n}$.

Thus, for $x \notin F_n$,
$$\limsup_{t\to 0} |\Phi_t * v(x) - u(x)| \le \widetilde{V^{(n)}}(x) + \sum_{i=i_n}^\infty |u_i(x)| < 2^{-n+1}.$$
Consequently, for $x \notin \cup_{n=m}^\infty F_n$, where m is arbitrary,
$$\limsup_{t\to 0} |\Phi_t * v(x) - u(x)| = 0.$$

But it is easily seen that $\mathrm{cap}(\cup_{n=m}^\infty F_n) < C(\sum_{n=m}^\infty 2^{-n\kappa})^{1/\kappa}$, which is arbitrarily small, and the claim (i) follows. $\square$

PROOF OF THEOREM 2.3.2. Suppose that f satisfies the assumptions of the theorem. The distribution f has a representation $f = \sum_{i=0}^\infty f_i$, where
$$f_i = \sum_{k\in\mathbf{Z}^N} s_{i,k} a_{i,k}, \quad a_{i,k} \perp \mathfrak{P}_{L-1}, \quad \text{and} \quad a_{i,k} \in C_0^M(Q_{i,k}(3)),$$
for some $M > l$. By Theorem 1.1.14 there is a function $\tilde f \in BL^l_{p,\theta}$ such that $\tilde f(x) = \sum_{i=0}^\infty f_i(x)$ with convergence in $L_{r,\mathrm{loc}}$ and pointwise a.e. By Theorem 2.1.4 $\tilde f$ has derivatives satisfying $D^\beta \tilde f(x) = \sum_{i=0}^\infty D^\beta f_i(x)$ for $|\beta| \le l$ q.e. with respect to the proper capacities. By Lemma 2.3.1 $D^\beta \tilde f(x) = \lim_{t\to 0} D^\beta f * \Phi_t(x)$ for $|\beta| \le l$ q.e. for any $\Phi \in \mathcal{S}$ such that $\int \Phi\, dx = 1$. Thus, by the assumptions $\sum_{i=0}^\infty D^\beta f_i(x) = 0$ for all $|\beta| \le l$ q.e. on A with respect to the proper capacities, i.e., the conclusion of Corollary 2.1.6 applies, which is the essential hypothesis of Theorem 2.2.1.

It follows that Theorem 2.2.1 can be applied to $\tilde f$. By the proof of that theorem there is a function η_ε such that $0 \le \eta_\varepsilon \le 1$, $\eta_\varepsilon = 1$ on an open neighborhood Ω of A, and $\eta_\varepsilon \tilde f = \sum_{i=0}^\infty u_i$, where $\{u_i\}_{i=0}^\infty$ satisfies (2.2.10), and
$$\|\eta_\varepsilon \tilde f\|_{BL^l_{p,\theta}} \le \|\{2^{il} u_i\}_{i=0}^\infty\|_{l_\theta(L_p)} < \varepsilon.$$
By Theorem 2.3.3 there is $v = \sum_{i=0}^\infty v_i$, such that $v \in B^l_{p,\theta}$,
$$\|v\|_{B^l_{p,\theta}} \le C\|\{2^{il} u_i\}_{i=0}^\infty\|_{l_\theta(L_p)} < C\varepsilon,$$
and $v = f$ on Ω in the sense of distributions, i.e., $\langle f - v, \varphi\rangle = 0$ for all $\varphi \in C_0^\infty(\Omega)$. It follows that $f_\varepsilon = f - v$ satisfies all requirements of the theorem. $\square$

2.4. Invariant Subspaces and a Theorem of Whitney

This section is devoted to a generalization of H. Whitney's well-known theorem (see [57]) characterizing the closed ideals in $C^s(\mathbf{R}^N)$. In the special case of $L_p^\lambda(\mathbf{R}^N)$, $\lambda > 0$, $1 < p < \infty$, the result was given with detailed proofs in [4], Section 10.2.

It is necessary to introduce some notation. Let $\mathfrak{P}$ denote the ring of all polynomials in N real variables. We denote by $\mathfrak{I}_s$, $s \in \mathbf{N}_0$, the ideal

$$\mathfrak{I}_s = \{\, p \in \mathfrak{P} : p(x) = \sum_{|\sigma|>s} a_\sigma x^\sigma \,\}.$$

The ring of residue classes (the quotient ring) mod $\mathfrak{I}_s$ is denoted $\mathfrak{P}/\mathfrak{I}_s$. We denote elements in $\mathfrak{P}_s$ by the same letters as their images in $\mathfrak{P}/\mathfrak{I}_s$ under the natural mapping.

For each $y \in \mathbf{R}^N$ the Taylor formula defines a ring homomorhism,

$$P_y^s : C^s \to \mathfrak{P}/\mathfrak{I}_s,$$

by

$$P_y^s f(x) = \sum_{|\sigma|\le s} D^\sigma f(y) \frac{x^\sigma}{\sigma!},$$

and the image of a closed ideal $\mathfrak{M}$ in C^s is an ideal in $\mathfrak{P}/\mathfrak{I}_s$, called the local ideal, $\mathfrak{M}_y$, of $\mathfrak{M}$ at y.

Whitney's result is the following.

THEOREM 2.4.1 (H. Whitney [57]). *Let $\mathfrak{M}$ be a closed ideal in $C^s(\mathbf{R}^N)$, $s \in \mathbf{N}_0$, and let $g \in C^s(\mathbf{R}^N)$. Then $g \in \mathfrak{M}$ if and only if $P_y^s g \in \mathfrak{M}_y$ for all $y \in \mathbf{R}^N$.*

The problem makes sense also in the spaces $YL(E)$ and $Y(E)$, although these are not in general algebras. They are, however, always invariant under multiplication by functions in C_0^∞. We shall characterize the closed subspaces of $YL(E)$, which are invariant under multiplication by C_0^∞ functions, i.e., we shall characterize the closed C_0^∞-submodules of $YL(E)$.

For $Y(E)$ the problem remains open. For the homogeneous spaces $\dot{Y}L(\dot{E})$ and $\dot{Y}(\dot{E})$ the problem cannot be posed, since their elements are equivalence classes which are not closed under multiplication by smooth functions.

Let $E \in S(\varepsilon_+, \varepsilon_-, r)$, where ε_+, ε_-, $r > 0$, and suppose that $f \in YL(E)$. Then the mapping $P_y^s : f \mapsto P_y^s f \in \mathfrak{P}/\mathfrak{I}_s$ is well defined for q.e. $y \in \mathbf{R}^N$ with respect to $\mathrm{cap}(\,\cdot\,, s, E)$ if $P_y^s f$ is the L_r-differential of order s at y of f. See Definition 2.1.3 and Theorem 2.1.4.

It is easily seen that P_y^s is a homomorphism in the sense that if f has an s-differential at y, then φf has an s-differential at y for all $\varphi \in C_0^\infty$, and

(2.4.1) $$P_y^s \varphi \, P_y^s f = P_y^s(\varphi f) \pmod{\mathfrak{I}_s}.$$

The main result is the following theorem, which contains Theorem 2.2.1 as a special case.

THEOREM 2.4.2. *Let $E \in S(\varepsilon_+, \varepsilon_-, r)$, where ε_+, ε_-, $r > 0$, and suppose that E satisfies the condition (2.2.1). Let $\mathfrak{M}$ be a closed C_0^∞-submodule of $YL(E)$. Let M be any integer greater than or equal to ε_-. Then there is a family of ideals $\{\,\mathfrak{E}_y \subset \mathfrak{P}/\mathfrak{I}_M : y \in \mathbf{R}^N\,\}$ with the property that a function f_0 in $YL(E)$ belongs to $\mathfrak{M}$ if and only if for all integers s, $0 \le s \le \varepsilon_-$,*

(2.4.2) $$P_y^s f_0 \in \mathfrak{E}_y \pmod{\mathfrak{I}_s} \quad \text{for } \mathrm{cap}(\,\cdot\,, s, E)\text{-q.e. } y \in \mathbf{R}^N.$$

2.4. INVARIANT SUBSPACES AND A THEOREM OF WHITNEY

Theorem 2.4.2 is an easy consequence of the following result, which is of independent interest.

We denote by $\mathcal{L}(V)$ the linear hull of a subset V of a vector space, i.e. $\mathcal{L}(V)$ is the set of finite linear combinations of elements in V.

THEOREM 2.4.3. *Let E be as in the previous theorem, and let $f_i \in YL(E)$, $i \in \mathbf{N}_0$. Then the following conditions are equivalent:*

(i) *For all integers s, $0 \leq s \leq \varepsilon_-$, and for $\mathrm{cap}(\,\cdot\,, s, E)$-q.e. $y \in \mathbf{R}^N$, the polynomial $P_y^s f_0$ belongs $(\mathrm{mod}\ \mathfrak{I}_s)$ to the ideal in $\mathfrak{P}/\mathfrak{I}_s$ generated by the polynomials $\{P_y^s f_i : i = 1, 2, \ldots\}$.*

(ii) *The function f_0 belongs to the closed C_0^∞-module generated by $\{f_i\}_1^\infty$, i.e., for any $\varepsilon > 0$ there is a finite subset Q of $\mathbf{N}$, and functions $\varphi_i \in C_0^\infty$, $i \in Q$, such that*

$$\left\| f_0 - \sum_{i \in Q} \varphi_i f_i \right\|_{YL(E)} < \varepsilon.$$

The proof of this theorem depends on the same technique as the proof of Theorem 2.2.1, and is quite long. It is omitted, since the proof given in [**4**] applies with the modifications introduced in proving Theorem 2.2.1.

CHAPTER 3

Luzin Type Theorems

3.1. Luzin Approximation of Functions

The main result of this section is Theorem 3.1.2, which is a theorem of so called Luzin type for functions in the spaces $YL(E)$.

In order to formulate the result we need the capacities $\operatorname{cap}(\,\cdot\,, s, E)$ associated to a space E belonging to $S(\varepsilon_+, \varepsilon_-, r)$, which were defined in Definition 2.1.1.

We also define "Hausdorff capacities" associated to the same spaces. Such capacities were introduced in [**34**]; see also [**36**].

DEFINITION 3.1.1. Let $s \geq 0$, $\varepsilon_+, \varepsilon_- \in \mathbf{R}$, $0 < r \leq \infty$, let $E \in S(\varepsilon_+, \varepsilon_-, r)$, and let $A \subset \mathbf{R}^N$. Consider all index sets $\Gamma \subset \mathbf{N}_0 \times \mathbf{Z}^N$ such that $A \subset \cup_{(i,k)\in\Gamma} Q_{i,k}$. Set $\Gamma(i) = \{\, k : (i,k) \in \Gamma \,\}$, and

$$g_i(x) = 2^{-is} \sum_{k \in \Gamma(i)} \chi_{i,k}(x).$$

Then we define

$$\Lambda(A, s, E) = \inf_\Gamma \|\{g_i\}_{i=0}^\infty\|_E,$$

taking the infimum over all such Γ.

REMARK. Clearly $\operatorname{cap}(A, s, E) \leq \Lambda(A, s, E)$ for all sets A. Note that Lemma 2.1.5(v) implies that $\operatorname{cap}(A, s, E) = 0$ for all sets A if $s > \varepsilon_-$.

REMARK. If $E = L_p(l_\theta^\lambda)$, which corresponds to the space $Y(E) = F_{p,\theta}^\lambda$, it is easily seen that $\Lambda(\,\cdot\,, s, E)^p$, $0 \leq s \leq \lambda$, for any θ, $0 < \theta \leq \infty$, is equivalent to the "ordinary" Hausdorff capacity, which in [**4**] (see p. 132) is denoted by $\Lambda_{N-(m-s)p}^{(\infty)}(\,\cdot\,)$. It is interesting to note that this independence of θ is also true for the corresponding capacities $\operatorname{cap}(\,\cdot\,, s, L_p(l_\theta^\lambda))$, see [**27**] and [**33**].

We shall use the Besov spaces $B_{\infty,\infty}^s$, which in our setting are $Y(E)$ with $E = L_\infty(l_\infty^s)$. Recall that when s is not an integer, say $s = k + r$ with $k \in \mathbf{N}_0$ and $0 < r < 1$, then $B_{\infty,\infty}^s$ can be identified with the space $C^{k,r}$ of k times continuously differentiable functions with r-Hölder continuous derivatives of order k. For $s \in \mathbf{N}$, however, $C^{s-1,1} = W_\infty^s$ is strictly contained in $B_{\infty,\infty}^s$. In particular, $B_{\infty,\infty}^1$ is the well-known Zygmund class. We denote the closure of smooth functions in $B_{\infty,\infty}^s$ by $B_{\infty,\infty,0}^s$. This corresponds to replacing l_∞ by the space c_0 of sequences converging to 0 in the definition of the space E.

THEOREM 3.1.2. Let ε_+, ε_-, $r > 0$, $0 < p \leq \infty$, $\frac{N}{p} > \frac{N}{r} - \varepsilon_+$, and $E \in S(\varepsilon_+, \varepsilon_-, r)$.

(a) If $s \in \mathbf{N}_0$, $0 \leq s \leq \varepsilon_-$, then for any $f \in YL(E)$ and any $\varepsilon > 0$ there is an open set G with $\operatorname{cap}(G, s, E) < \varepsilon$ and a function $g \in W_\infty^s \cap YL(E)$ such that $f = g$

on G^c. Moreover,

(3.1.1) $$\|g\|_{W^s_\infty} \leq C\varepsilon^{-1}\|f\|_{YL(E)},$$

(3.1.2) $$\|g\|_{YL(E)} \leq C\|f\|_{YL(E)},$$

and for any $x \in G^c$,

(3.1.3) $$\sup_{0<a\leq 1}\left(\frac{1}{a^{N+sp}}\int_{B(x,a)}|f-g|^p\,dy\right)^{1/p} \leq C\varepsilon^{-1}\|f\|_{YL(E)},$$

and

(3.1.4) $$\lim_{a\to 0}\frac{1}{a^{N+sp}}\int_{B(x,a)}|f-g|^p\,dy = 0.$$

If $s \geq 1$ the function g has an L_∞-differential of order s at each point, and if $s = 0$ each point is a Lebesgue L_q-point of g for any $q < \infty$.

If, in addition, it is known that $C^s \cap YL(E)$ is dense in $YL(E)$, then one can also require that $g \in C^s$. In this case $\|f - g\|_{YL(E)}$ can also be made arbitrarily small.

(b) If $s \in \mathbf{R}$, $0 < s \leq \varepsilon_-$, then for any $f \in YL(E)$ and any $\varepsilon > 0$ there is an open set G with $\Lambda(G, s, E) < \varepsilon$ and a function $g \in B^s_{\infty,\infty} \cap YL(E)$ such that $f = g$ on G^c. Moreover,

(3.1.5) $$\|g\|_{B^s_{\infty,\infty}} \leq C\varepsilon^{-1}\|f\|_{YL(E)},$$

(3.1.6) $$\|g\|_{YL(E)} \leq C\|f\|_{YL(E)},$$

and (3.1.3) holds for all $x \in G^c$ and $0 < a \leq 1$.

If, in addition, $B^s_{\infty,\infty,0} \cap YL(E)$ is dense in $YL(E)$, one can also require that $g \in B^s_{\infty,\infty,0}$, and then (3.1.4) holds for any $x \in G^c$. In this case $\|f - g\|_{YL(E)}$ can also be made arbitrarily small.

REMARK. This kind of result, which generalizes the well-known Luzin theorem, has a rather long history. See e.g. Yu. A. Brudnyi [12] (Theorem 3.4), the book by W. P. Ziemer [60] and references given there, and the more recent papers by B. Bojarski, P. Hajłasz, and P. Strzelecki [8], and by D. Swanson [52]. The known results deal mainly ([12] is an exception) with Bessel potential spaces L^k_p (Sobolev spaces W^k_p if $k \in \mathbf{N}$) for $1 < p < \infty$, in which case the condition $\text{cap}(G, s, E) < \varepsilon$ becomes $C_{k-s,p}(G) < \varepsilon^p$, and $\Lambda(G, s, E) < \varepsilon^p$ becomes $\Lambda^{(\infty)}_{N-(k-s)p}(G) < \varepsilon$.

In addition to the extension to more general spaces, Theorem 3.1.2 improves earlier results in two ways. Firstly, our approximating function g inherits the regularity of the space $YL(E)$ and approximates f in the sense of that space, whereas in [8] and [52] it approximates f in the sense of L^{s+1}_p, $s + 1 \leq k$.

Secondly, in part (b) the exceptional set G was previously known only to have small $C_{k-s,p}$-capacity. It follows from known comparison theorems that our result is strictly stronger, see e.g. Theorem 5.4.2 in [4].

Part of the results given here were announced by Netrusov in 1987, but no proofs have been published. See [4], Section 6.5.3 for more details.

Finally, we take the opportunity to make a correction. In [4], Section 6.5.2, a result of Michael and Ziemer was incorrectly quoted: $\|f - f_\varepsilon\|_{\alpha,p}$ should have been $\|f - f_\varepsilon\|_{k,p}$.

The proof of Theorem 3.1.2 depends on the following two propositions.

3.1. LUZIN APPROXIMATION OF FUNCTIONS

PROPOSITION 3.1.3. *Let $f \in L_{p,\mathrm{loc}}$, $0 < p < \infty$, $M \in \mathbf{N}$, and $0 < s < M$. Let G be defined by*

$$(3.1.7) \qquad G = \left\{ x \in \mathbf{R}^N : \|f\|_{L_p(B(x,1))} + \sup_{0 < a \leq 1} \frac{\mathcal{E}_M(f, B(x,a), L_p)}{a^{s+N/p}} > \lambda \right\}.$$

Then there is a function $g \in B^s_{\infty,\infty}(\mathbf{R}^N)$ with $\|g\|_{B^s_{\infty,\infty}} \leq C\lambda$, which has the following properties:

(i) *For any $x \in G^c$ we have*

$$(3.1.8) \qquad \sup_{0 < a \leq 1} \left(\frac{1}{a^{N+sp}} \int_{B(x,a)} |f-g|^p \, dy \right)^{1/p} \leq C\lambda.$$

(ii) *There are constants C_1 and C_2 such that*

$$(3.1.9) \qquad \mathcal{E}_M(g, B(x,a), L_p) \leq C_1 \mathcal{E}_M(f, B(x, C_2 a), L_p)$$

for all x and a, $0 < a \leq 1$, such that $B(x, 4a) \cap G^c \neq \emptyset$.

(iii) *For all x and a, $0 < a \leq 1$, such that $B(x, 4a) \subset G$,*

$$(3.1.10) \quad \mathcal{E}_M(g, B(x,a), L_p)$$
$$\leq C \left(\frac{a}{\mathrm{dist}(x, \partial G)} \right)^{M+N/p} \mathcal{E}_M(f, B(x, \tfrac{1}{2} \mathrm{dist}(x, \partial G)), L_p).$$

PROOF. We fix a $\lambda > 0$. The corresponding set G is clearly open. Let $\mathcal{Q} = \{Q\}$ be a Whitney covering of G, i.e., $\mathcal{Q}$ is a disjoint subfamily of the family of dyadic cells $\{Q_{i,k}\}$ such that $G = \cup_{Q \in \mathcal{Q}} Q$, as in e.g. E. M. Stein [48], Ch. VI.1, pp. 167–170. By further subdividing the cubes we can assume that

$$(3.1.11) \qquad \mathrm{diam}\, Q \leq \tfrac{1}{8} \mathrm{dist}(Q, G^c) \leq 4 \,\mathrm{diam}\, Q \quad \text{for all } Q \in \mathcal{Q}.$$

Let $1 < \rho < \tfrac{5}{4}$, and denote the expanded cubes $Q(\rho)$ by Q^*. Then ([48], Ch. VI.1.3) there is a number C_N (which can be chosen as 12^N) such that each point in G is contained in at most C_N of the cubes Q^*, $Q \in \mathcal{Q}$. Furthermore, there is a partition of unity $\{\varphi_Q\}_{Q \in \mathcal{Q}}$, $\sum_{Q \in \mathcal{Q}} \varphi_Q = \chi_G$, such that $\varphi_Q \in C^M$, $\mathrm{supp}\, \varphi_Q \subset Q^*$, and

$$(3.1.12) \qquad \max_x |D^\gamma \varphi_Q(x)| \leq C (\mathrm{diam}\, Q)^{-|\gamma|}$$

for all multi-indices γ with $|\gamma| \leq M$.

For each $Q \in \mathcal{Q}$ we choose a polynomial $\pi_Q \in \mathfrak{P}_{M-1}$ such that

$$\|f - \pi_Q\|_{L_p(Q^*)} \leq 2 \mathcal{E}_M(f, Q^*, L_p).$$

We define the function g by

$$g(x) = f(x), \quad x \notin G;$$
$$g(x) = \sum_{Q \in \mathcal{Q}} \varphi_Q(x) \pi_Q(x), \quad x \in G,$$

or equivalently

$$(3.1.13) \qquad g = f + \sum_{Q \in \mathcal{Q}} \varphi_Q (\pi_Q - f).$$

We first prove (3.1.8) and (3.1.9). Let $0 < a \leq 1$, and let x be a point such that $x \in G^c$, or $x \in G$ and $\text{dist}(x, \partial G) \leq 4a$. Choose a constant C_1 so that $Q^* \subset B(x, C_1 a)$ if Q^* intersects $B(x, a)$, and a polynomial $\pi \in \mathfrak{P}_{M-1}$ such that

$$\|f - \pi\|_{L_p(B(x, C_1 a))} \leq 2\mathcal{E}_M(f, B(x, C_1 a), L_p).$$

We shall estimate

$$\|g - f\|_{L_p(B(x,a))} = \left\|\sum_{Q \in \mathcal{Q}} \varphi_Q(\pi_Q - f)\right\|_{L_p(B(x,a))}.$$

By the properties of the Whitney covering we obtain

$$\int_{B(x,a)} \left|\sum_{Q \in \mathcal{Q}} \varphi_Q(\pi_Q - f)\right|^p dy \leq C \int_{B(x,a)} \left|\sum_{Q \in \mathcal{Q}} \chi(Q^*)|\pi_Q - f|\right|^p dy$$

$$\leq C \int_{B(x,a)} \sum_{Q \in \mathcal{Q}} \chi(Q^*)|\pi_Q - f|^p \, dy \leq C \sum_{Q^* \cap B(x,a) \neq \emptyset} \int_{Q^*} |\pi_Q - f|^p \, dy$$

$$\leq C \sum_{Q^* \cap B(x,a) \neq \emptyset} \int_{Q^*} |\pi - f|^p \, dy \leq C \int_{B(x, C_1 a)} |\pi - f|^p \sum_{Q \in \mathcal{Q}} \chi(Q^*) \, dy$$

$$\leq C \int_{B(x, C_1 a)} |\pi - f|^p \, dy,$$

and thus

(3.1.14) $$\int_{B(x,a)} |f - g|^p \, dy \leq C \int_{B(x, C_1 a)} |f - \pi|^p \, dy.$$

If $x \in G^c$ this implies (3.1.8).

The inequality (3.1.9) now follows immediately from (3.1.7), since $g - \pi = f - \pi + \sum_{Q \in \mathcal{Q}} \varphi_Q(\pi_Q - f)$.

In order to prove (3.1.10) we let $0 < a \leq 1$, and let $x \in G$ satisfy $\text{dist}(x, \partial G) > 4a$. We now choose $\pi \in \mathfrak{P}_{M-1}$ so that

$$\|f - \pi\|_{L_p(B(x, \frac{1}{2} \text{dist}(x, \partial G)))} \leq 2\mathcal{E}_M(f, \frac{1}{2} \text{dist}(x, \partial G)), L_p).$$

Then $g - \pi = \sum_{Q \in \mathcal{Q}} \varphi_Q(\pi_Q - \pi)$ in $B(x, a)$. We have

$$\mathcal{E}_M(g, B(x, a), L_p) = \mathcal{E}_M(g - \pi, B(x, a), L_p)$$

$$\leq C \left\|\sum_{Q \in \mathcal{Q}} (\varphi_Q(\pi_Q - \pi) - P_x^{M-1}(\varphi_Q(\pi_Q - \pi)))\right\|_{L_p(B(x,a))},$$

where P_x^{M-1} denotes the Taylor polynomial of degree $M - 1$ at x.

Since the number of cubes $Q \in \mathcal{Q}$ such that Q^* intersects $B(x, a)$ is bounded, it is enough to estimate the remainder term in the Taylor expansion of one of the terms. For a cube Q^* that intersects $B(x, a)$ we have

$$\|\varphi_Q(\pi_Q - \pi) - P_x^{M-1}(\varphi_Q(\pi_Q - \pi))\|_{L_p(B(x,a))}$$
$$\leq C a^{M+N/p} \max_{|\gamma|=M} \|D^\gamma(\varphi_Q(\pi_Q - \pi))\|_{L_\infty(Q^*)}.$$

By (3.1.12) and (3.1.11) we have

$$\|D^\alpha \varphi_Q\|_{L_\infty} \leq C \, \text{dist}(x, \partial G)^{-|\alpha|} \quad \text{for } |\alpha| \leq M.$$

By the equivalence of (quasi-)norms on finite dimensional spaces (Lemma 1.2.2) there is a constant $C = C(M,p)$ such that

$$\|D^\beta(\pi_Q - \pi)\|_{L_\infty(Q^*)} \le C|Q|^{-(|\beta|/N+1/p)}\|\pi_Q - \pi\|_{L_p(Q^*)} \quad \text{for } |\beta| \le M.$$

Observing that the inequality (3.1.11) ensures that $Q^* \subset B(x, \frac{1}{2}\operatorname{dist}(x, \partial G))$ if Q^* intersects $B(x, a)$ and $a < \frac{1}{4}\operatorname{dist}(x, \partial G)$, it follows from the choice of π and π_Q that

$$\|\pi_Q - \pi\|_{L_p(Q^*)} \le C\mathcal{E}_M(f, B(x, \tfrac{1}{2}\operatorname{dist}(x, \partial G)), L_p).$$

Consequently, by the Leibniz rule and (3.1.12)

$$\max_{|\gamma|=M} \|D^\gamma(\varphi_Q(\pi_Q - \pi))\|_{L_\infty(Q^*)} \le \frac{C\mathcal{E}_M(f, B(x, \tfrac{1}{2}\operatorname{dist}(x, \partial G)), L_p)}{\operatorname{dist}(x, \partial G)^{M+N/p}},$$

which proves (3.1.10).

We finally prove the inequality $\|g\|_{B^s_{\infty,\infty}} \le C\lambda$. By (3.1.7) and (3.1.9) we have $\sup_{0 < a \le 1} a^{-s-N/p}\mathcal{E}_M(g, B(x, a), L_p) \le C\lambda$ for $x \in G^c$.

If $x \in G$ we let $x^* \in \partial G$ be a point such that $|x - x^*| = \operatorname{dist}(x, \partial G)$, and observe that $B(x, a) \subset B(x^*, 5a)$ if $\frac{1}{4}\operatorname{dist}(x, \partial G) \le a \le 1$. It follows again from (3.1.7) and (3.1.9) that

$$(3.1.15) \qquad a^{-s-N/p}\mathcal{E}_M(g, B(x, a), L_p) \le a^{-s-N/p}\mathcal{E}_M(g, B(x^*, 5a), L_p) \le C\lambda$$

for these a.

If $x \in G$ and $0 < a < \frac{1}{4}\operatorname{dist}(x, \partial G) \le 1$ it follows from (3.1.9), (3.1.10) and (3.1.7) that

$$a^{-s-N/p}\mathcal{E}_M(g, B(x, a), L_p) \le \frac{Ca^{M-s}}{\operatorname{dist}(x, \partial G)^{M+N/p}}\mathcal{E}_M(g, B(x^*, \tfrac{3}{2}\operatorname{dist}(x, \partial G)), L_p)$$

$$\le C\lambda \frac{a^{M-s}}{\operatorname{dist}(x, \partial G)^{M-s}} \le C\lambda.$$

Consequently

$$\sup_{0 < a \le 1} a^{-s-N/p}\mathcal{E}_M(g, B(x, a), L_p) \le C\lambda$$

for all x.

It follows from the well-known Morrey–Campanato lemma (see e.g. Triebel [54], Th.1.7.3), and also from Theorem 1.1.14(iii), that $g \in B^s_{\infty,\infty}$ with $\|g\|_{B^s_{\infty,\infty}} \le C\lambda$. $\square$

PROPOSITION 3.1.4. *Let $f \in L_{p,\mathrm{loc}}$, $0 < p < \infty$, and let s and M be integers, $0 \le s \le M-1$. Let $G \subset \mathbf{R}^N$ be the set of points x such that*

$$(3.1.16) \qquad \|f\|_{L_p(B(x,1))} + \sum_{i=1}^\infty 2^{i(s+N/p)}\mathcal{E}_M(f, B(x, 2^{-i}), L_p) > \lambda.$$

Then there exists a function $g \in W^s_\infty(\mathbf{R}^N)$ with $\|g\|_{W^s_\infty} \le C\lambda$, such that (i), (ii) and (iii) in Proposition 3.1.3 hold true, and such that (3.1.8) and (3.1.4) are satisfied for all $x \in G^c$. Moreover, if $s \ge 1$, g has an L_∞-differential of order s at all x, and if $s = 0$ every point $x \in \mathbf{R}^N$ is an L_q-Lebesgue point for g for all $q < \infty$.

PROOF. Let G be an open set such that (3.1.16) is satisfied, and let g be constructed as in the preceding proof.

Then there is a C such that

$$\|g\|_{L_p(B(x,1))} + \sum_{i=1}^{\infty} 2^{i(s+N/p)} \mathcal{E}_M(g, B(x, 2^{-i}), L_p) \leq C\lambda \tag{3.1.17}$$

for all x. In fact, for $x \in G^c$ this is immediate from (3.1.9), and for $x \in G$ we have by (3.1.10) that

$$\mathcal{E}_M(g, B(x, 2^{-i}), L_p)$$
$$\leq C 2^{-i(M+N/p)} \operatorname{dist}(x, \partial G)^{-M} \mathcal{E}_M(f, B(x, \tfrac{1}{2}\operatorname{dist}(x, \partial G)), L_p)$$

for $2^{-i} < \tfrac{1}{4}\operatorname{dist}(x, \partial G))$. Thus, if we define an integer i_0 so that

$$2^{-i_0} < \tfrac{1}{4}\operatorname{dist}(x, \partial G)) \leq 2^{-i_0+1},$$

we obtain as in (3.1.15)

$$\sum_{i=1}^{i_0-1} 2^{i(s+N/p)} \mathcal{E}_M(g, B(x, 2^{-i}), L_p)$$
$$\leq C \sum_{i=1}^{i_0-1} 2^{i(s+N/p)} \mathcal{E}_M(g, B(x^*, 5 \cdot 2^{-i}), L_p) \leq C\lambda,$$

and by (3.1.9)

$$\sum_{i=i_0}^{\infty} 2^{i(s+N/p)} \mathcal{E}_M(g, B(x, 2^{-i}), L_p)$$
$$\leq C \operatorname{dist}(x, \partial G)^{-M-N/p} \mathcal{E}_M(f, B(x, \tfrac{1}{2}\operatorname{dist}(x, \partial G)), L_p) \sum_{i=i_0}^{\infty} 2^{i(s-M)}$$
$$\leq C \operatorname{dist}(x, \partial G)^{-M-N/p} \mathcal{E}_M(f, B(x, \tfrac{1}{2}\operatorname{dist}(x, \partial G)), L_p) 2^{i_0(s-M)}$$
$$\leq C \operatorname{dist}(x, \partial G)^{-s-N/p} \mathcal{E}_M(f, B(x^*, \tfrac{3}{2}\operatorname{dist}(x, \partial G)), L_p) \leq C\lambda.$$

Then the relations (3.1.8) and (3.1.4) follow immediately from (3.1.14), (3.1.16) and (3.1.17) as before.

We shall now prove that (3.1.17) implies that $g \in W_\infty^s$ with $\|g\|_{W_\infty^s} < C\lambda$. We apply the proof of the implication (iii*) $\Rightarrow$ (iv) in the proof of Theorem 1.1.14. It follows from (1.3.3) that g has an atomic representation

$$g(x) = \sum_{i=i_1}^{\infty} g_i(x) = \sum_{i=i_1}^{\infty} \sum_{k \in \mathbf{Z}^N} s_{i,k} a_{i,k}(x),$$

where $\operatorname{supp} a_{i,k} \subset Q_{i,k}(3)$, $\|D^\beta a_{i,k}\|_{L_\infty} \leq C 2^{i|\beta|}$ for $|\beta| \leq M$, and i_1 can be chosen so large that

$$|s_{i,k}| \leq C 2^{iN/p} \mathcal{E}_M(g, B(y, 2^{-i+i_1}), L_p)$$

for all $y \in \operatorname{supp} a_{i,k}$. It follows from (3.1.17) that

$$\sum_{i=i_1}^{\infty} 2^{i(s-|\beta|)} |D^\beta g_i(y)| < C\lambda \quad \text{for all } y \text{ and all } |\beta| \leq s.$$

If $s \geq 1$ this implies that $g \in C^{s-1}$, and that the derivatives of order $s-1$ are Lipschitz continuous, i.e., $g \in W_\infty^s$, with $\|g\|_{W_\infty^s} < C\lambda$. If $s = 0$ we obtain $g \in L_\infty$ and $\|g\|_{L_\infty} < C\lambda$.

We prove that for $1 \leq s \leq M-1$ the function g has an L_∞-differential of order s at each point x. The Taylor polynomial $P_x^s g \in \mathfrak{P}_s$ defined by

$$P_x^s g(y) = \sum_{i=0}^\infty P_x^s g_i(y) = \sum_{|\beta| \leq s} \frac{(y-x)^\beta}{\beta!} \sum_{i=0}^\infty D^\beta g_i(x)$$

is well defined for all x. (Cf. the proof of Theorem 2.1.4.)

We fix x, we assume without loss of generality that $x = 0$, and we denote

$$2^{iN/p} \mathcal{E}_M(g, B(0, 2^{-i}), L_p) = E_i.$$

We claim that

$$\lim_{i \to \infty} 2^{is} \|g - P_0^s g\|_{L_\infty(B(0, 2^{-i}))} = 0.$$

We choose an $\varepsilon > 0$, and we fix j so that

(3.1.18) $$\sum_{i=j}^\infty 2^{is} E_{i-i_1} < \varepsilon.$$

We have

$$|g - P_0^s g| \leq \sum_{i=i_1}^j |g_i - P_0^s g_i| + \sum_{i=j+1}^\infty |g_i| + \sum_{i=j+1}^\infty |P_0^s g_i|.$$

For $y \in B(0, 2^{-j})$ we obtain for the first and third terms, respectively,

$$\sum_{i=i_1}^j |g_i(y) - P_0^s g_i(y)| \leq C 2^{-j(s+1)} \sum_{i=i_1}^j \sum_{|\beta|=s+1} \max_{\xi \in B(0, 2^{-j})} |D^\beta g_i(\xi)|$$

$$\leq C 2^{-j(s+1)} \sum_{i=i_1}^j 2^{i(s+1)} E_{i-i_1} \leq C 2^{-js} \sum_{i=i_1}^j 2^{i-j} 2^{is} E_{i-i_1},$$

and

$$\sum_{i=j+1}^\infty |P_0^s g_i(y)| \leq C \sum_{l=0}^s 2^{-jl} \sum_{i=j+1}^\infty 2^{il} E_{i-i_1} \leq C 2^{-js} \sum_{i=j+1}^\infty 2^{is} E_{i-i_1} \leq C 2^{-js} \varepsilon.$$

In order to estimate the second term we let n be a positive integer, to be chosen shortly.

Then for $y \in B(0, 2^{-j})$

$$\sum_{i=j+1}^{j+n} |g_i(y)| \leq C \sum_{i=j+1}^{j+n} 2^{iN/p} \mathcal{E}_M(g, B(y, 2^{-i+i_1}), L_p)$$

$$\leq C \mathcal{E}_M(g, B(0, 2^{-j+i_1}), L_p) \sum_{i=j+1}^{j+n} 2^{iN/p}$$

$$\leq C \mathcal{E}_M(g, B(0, 2^{-j+i_1}), L_p) 2^{(j+n)N/p}$$

$$= C 2^{nN/p} E_{j-i_1} < C 2^{nN/p} 2^{-js} \varepsilon.$$

On the other hand

$$|g_i(y)| \leq C\lambda 2^{-is}$$

for all $i \geq i_1$, and thus
$$\sum_{i=j+n+1}^{\infty} |g_i(y)| \leq C\lambda \sum_{i=j+n+1}^{\infty} 2^{-is} \leq C\lambda 2^{-(j+n)s}.$$

Adding these estimates we have
$$\sum_{i=j+1}^{\infty} |g_i(y)| \leq C 2^{-js}(2^{nN/p}\varepsilon + \lambda 2^{-ns}).$$

If we choose the integer n so that $2^{nN/p}\varepsilon$ is close to $\lambda 2^{-ns}$ we find that there is a ρ, $0 < \rho < 1$, such that
$$2^{nN/p}\varepsilon + \lambda 2^{-ns} < C\varepsilon^\rho .$$

Combining the estimates it finally follows that
$$2^{js}\|g - P_0^s g\|_{L_\infty(B(0,2^{-j}))} \leq C \sum_{i=i_1}^{j} 2^{i-j} 2^{is} E_i + C\varepsilon^\rho.$$

Thanks to the convergence of $\sum_{i=0}^{\infty} 2^{is} E_i$ this proves the claim if j is allowed to tend to infinity.

We note that this argument breaks down for $s = 0$. In fact, in this case the condition (3.1.17) does not imply that g is continuous. We obtain a counterexample by taking $g(x) = \sum_{i=1}^{\infty} \varphi(2^{2^i}(x - 2^{-i}))$ where $\varphi \in C_0(B(0, \frac{1}{4}))$.

Now let $s = 0$. In order to prove that each x is an L_q-Lebesgue point for g for all $q < \infty$ it is enough to prove that it is an L_p-Lebesgue point. In fact, if there is an $a_0 = a_0(x)$ so that
$$\lim_{i \to \infty} 2^{iN/p}\|g - a_0\|_{L_p(B(x, 2^{-i}))} = 0,$$

it follows that for all q such that $p \leq q < \infty$
$$\lim_{a \to 0} \frac{1}{a^N} \int_{B(x,a)} |g(y) - a_0(x)|^q \, dy \leq \lim_{a \to 0} \frac{C\lambda^{q-p}}{a^N} \int_{B(x,a)} |g(y) - a_0(x)|^p \, dy = 0.$$

We again fix $x = 0$, and for each $i \in \mathbf{N}_0$ we choose $\pi_i \in \mathfrak{P}_{M-1}$ so that $\pi_0 = \|g\|_{L_p(B(0,1))}$, and for $i \geq 1$
$$\|g - \pi_i\|_{L_p(B(0,2^{-i}))} \leq 2\mathcal{E}_M(g, B(0, 2^{-i}), L_p) = 2E_i.$$

Then
$$\|\pi_i - \pi_{i+1}\|_{L_p(B(0,2^{-i-1}))} \leq CE_i \quad \text{for } i \geq 1,$$
and if we write $\pi_i(y) = \sum_{|\alpha| \leq M-1} a_{i,\alpha} y^\alpha$, it follows from the equivalence of norms on finite dimensional spaces (Lemma 1.2.2) that

(3.1.19) $$|a_{i,\alpha} - a_{i+1,\alpha}| \leq C 2^{i|\alpha|} E_i$$

for $|\alpha| \leq M - 1$. For $\alpha = 0$ it follows from (3.1.17) with $s = 0$ that

(3.1.20) $$\sum_{i=0}^{\infty} |a_{i,0} - a_{i+1,0}| \leq C\lambda,$$

and that
$$\lim_{i \to \infty} a_{i,0} = a_{0,0} + \sum_{i=0}^{\infty} (a_{i+1,0} - a_{i,0}) = a_0$$

exists with

(3.1.21) $$|a_0| \leq C\lambda.$$

For $1 \leq |\alpha| \leq M-1$ the inequality (3.1.19) gives similarly

(3.1.22) $$|a_{j,\alpha}| \leq C \sum_{i=0}^{j} 2^{i|\alpha|} E_i.$$

We claim that (3.1.19) and (3.1.22) imply

(3.1.23) $$\lim_{j\to\infty} 2^{jN/p} \|g - a_0\|_{L_p(B(0,2^{-j}))} = 0 \ .$$

We have for each j

$$|g(y) - a_0| \leq |g(y) - \pi_j(y)| + |a_{j,0} - a_0| + \left| \sum\nolimits_{1 \leq |\alpha| \leq M-1} a_{j,\alpha} y^\alpha \right|.$$

We estimate the three terms separately. By the choice of π_j we already have

$$\lim_{j\to\infty} 2^{jN/p} \|g - \pi_j\|_{L_p(B(0,2^{-j}))} \leq C \lim_{j\to\infty} E_j = 0.$$

For the second term we have by (3.1.19)

$$|a_{j,0} - a_0| \leq \sum_{i=j}^{\infty} |a_{i,0} - a_{i+1,0}| \leq C \sum_{i=j}^{\infty} E_i,$$

and (3.1.18) gives the desired estimate.

In the third term (3.1.22) gives for $y \in B(0, 2^{-j})$

$$\left| \sum\nolimits_{1 \leq |\alpha| \leq M-1} a_{j,\alpha} y^\alpha \right| \leq C \sum_{i=0}^{j} \sum_{1 \leq |\alpha| \leq M-1} 2^{-j|\alpha|} 2^{i|\alpha|} E_i.$$

But it again follows from the convergence of $\sum_{j=0}^{\infty} E_j$, that for each multi-index α with $1 \leq |\alpha| \leq M-1$

$$\lim_{j\to\infty} \sum_{i=0}^{j} 2^{-(j-i)|\alpha|} E_i = 0 \ .$$

The desired estimate follows, and the claim (3.1.23) is proved. $\square$

PROOF OF THEOREM 3.1.2. Let $f \in YL(E)$, and let $M \geq [\varepsilon_-] + 1$. By Theorem 1.1.14(iii), the functions

$$h_0(x) = \|f\|_{L_p(B(x,1))}, \quad h_i(x) = 2^{iN/p} \mathcal{E}_M(f, B(x, 2^{-i}), L_p), \quad i = 1, 2, \ldots,$$

satisfy $\|\{h_i\}_{i=0}^{\infty}\|_E \leq C \|f\|_{YL(E)}$ for any p, $0 < p \leq \infty$, $\frac{N}{p} > \frac{N}{r} - \varepsilon_+$. Thus, for any $s \geq 0$ and any $\lambda > 0$, the set

$$G_s = \left\{ y : \sum\nolimits_{i=0}^{\infty} 2^{is} h_i(y) > \lambda \right\}$$

satisfies

$$\mathrm{cap}(G_s, s, E) < C\lambda^{-1} \|f\|_{YL(E)}.$$

Similarly, the set

$$H_s = \left\{ y : \sup_{i \in \mathbf{N}_0} 2^{is} h_i(y) > \lambda \right\}$$

satisfies

$$\Lambda(H_s, s, E) < C\lambda^{-1} \|f\|_{YL(E)}.$$

It follows from Proposition 3.1.3 that if $\varepsilon > 0$ is given, and if λ is chosen so that $\lambda = C\varepsilon^{-1}\|f\|_{YL(E)}$, then the function $g \in B^s_{\infty,\infty}$ constructed in the proposition differs from f on a set H_s satisfying $\Lambda(H_s, s, E) < \varepsilon$, and satisfies (3.1.3) and (3.1.5). Moreover, g satisfies (3.1.9) and (3.1.10), and this easily implies (3.1.6) by Theorem 1.1.14(iii).

If s is an integer, $s = 0, 1, \ldots, M - 1$, it follows in the same way from Proposition 3.1.4 that there is a function $g \in W^s_\infty$ which differs from f on a set G_s such that $\operatorname{cap}(G_s, s, E) < \varepsilon$, and satisfies all desired conditions.

We now assume that smooth functions are dense in $YL(E)$. We can assume that $f = \sum_{i=0}^\infty f^{(i)}$ where each $f^{(i)} \in C^M$ and $\|f^{(i)}\|_{YL(E)} < 4^{-i}\|f\|_{YL(E)}$. It is easily seen that if $f \in C^M$, then the corresponding function g constructed in Proposition 3.1.3 also belongs to C^M.

Thus for any $\varepsilon > 0$ and each $f^{(i)}$ there is a function $g^{(i)} \in C^s$ which differs from $f^{(i)}$ on a set $G^{(i)}$ such that $\operatorname{cap}(G^{(i)}, s, E) < \varepsilon 2^{-i}$ and satisfies
$$\|g^{(i)}\|_{YL(E)} < C\|f^{(i)}\|_{YL(E)} < C4^{-i}\|f\|_{YL(E)},$$
and
$$\|g^{(i)}\|_{C^s} < C\varepsilon^{-1}2^i\|f^{(i)}\|_{YL(E)} < C\varepsilon^{-1}2^{-i}\|f\|_{YL(E)}.$$
Moreover
$$\left(\frac{1}{a^{N+sp}}\int_{B(x,a)} |f^{(i)} - g^{(i)}|^p \, dy\right)^{1/p} \leq C\varepsilon^{-1}2^{-i}\|f\|_{YL(E)},$$
for any $x \in G^c$ and any a, $0 < a \leq 1$, and
$$\lim_{a \to 0} \frac{1}{a^{N+sp}}\int_{B(x,a)} |f^{(i)} - g^{(i)}|^p \, dy = 0$$
for each i.

If we set $g = \sum_{i=0}^\infty g^{(i)}$ and $G = \cup_{i=0}^\infty G^{(i)}$, it follows that $g \in YL(E) \cap C^s$,
$$\|g\|_{YL(E)} < C\|f\|_{YL(E)},$$
$$\|g\|_{C^s} < C\varepsilon^{-1}\|f\|_{YL(E)},$$
and g differs from f only on G. Moreover, if κ is the number defined in (1.1.2) it is easily seen that $\operatorname{cap}(G, s, E) < C\varepsilon(\sum_{i=0}^\infty 2^{-i\kappa})^{1/\kappa} = C\varepsilon$. The inequality (3.1.3) follows immediately from the quasi-Banach property of L_p. In order to prove (3.1.4) it is enough to choose an integer m so large that for a given $\delta > 0$
$$\sup_{0 < a \leq 1} \frac{1}{a^{N+sp}} \int_{B(x,a)} \left|\sum_{i=m}^\infty (f^{(i)} - g^{(i)})\right|^p dy \leq \delta,$$
and then observe that
$$\limsup_{a \to 0} \frac{1}{a^{N+sp}} \int_{B(x,a)} |f - g|^p \, dy$$
$$\leq C \limsup_{a \to 0} \frac{1}{a^{N+sp}} \int_{B(x,a)} \left|\sum_{i=m}^\infty (f^{(i)} - g^{(i)})\right|^p dy \leq C\delta.$$

Finally, if we write $f = f' + f''$, where $f' \in C^s$ and $\|f''\|_{YL(E)} < \varepsilon$, and apply the above to f'', it follows that there is a function $g'' \in C^s$ which differs from f'' only on a set G with $\operatorname{cap}(G, s, E) < \varepsilon$ such that $\|g''\|_{YL(E)} < C\varepsilon$. Thus, $g = f' + g''$ belongs to C^s, differs from f only on G, and $\|f - g\|_{YL(E)} < C\varepsilon$.

The last part of Theorem 3.1.2(b) is proved in the same way. $\square$

3.2. Luzin Approximation of Distributions

In this section we use Theorem 1.1.15 to prove a result analogous to Theorem 3.1.2 for distributions in the spaces $Y(E)$.

Let $f \in \mathcal{S}'$, and let $L, M \in \mathbf{N}_0$. Denote by $\mathcal{T}_M(f, B(x,a), L)$ the measurable function of x defined by

$$(3.2.1) \qquad \mathcal{T}_M(f, B(x,a), L) = \sup_\psi |\langle f, \psi \rangle|,$$

where the supremum is taken over all functions $\psi \in C_0^\infty(B(x,a))$ such that $\psi \perp \mathfrak{P}_{M-1}$ and $\|\psi\|_{C^L} \leq 1$.

As is shown by the following lemma, this quantity is a measure of the degree of local polynomial approximation to the distribution f.

LEMMA 3.2.1. *Let $L, M \in \mathbf{N}_0$. There is a constant C such that for all $f \in \mathcal{S}'$ and $0 < a \leq 1$*

$$(3.2.2) \quad \mathcal{T}_M(f, B(x,a), L) \leq \min_{\pi \in \mathfrak{P}_{M-1}} \mathcal{T}_0(f - \pi, B(x,a), L) \leq C \mathcal{T}_M(f, B(x,a), L).$$

PROOF. The left hand inequality is trivial. In fact, for any $\psi \in C_0^\infty(B(x,a))$ such that $\|\psi\|_{C^L} \leq 1$, and $\psi \perp \mathfrak{P}_{M-1}$, we have for any $\pi \in \mathfrak{P}_{M-1}$

$$|\langle f, \psi \rangle| = |\langle f - \pi, \psi \rangle| \leq \mathcal{T}_0(f - \pi, B(x,a), L).$$

In order to prove the right hand inequality we manufacture a polynomial $\pi_f \in \mathfrak{P}_{M-1}$ by means of dual bases constructed as in the proof of Lemma 1.4.2.

We assume without loss of generality that $x = 0$. Thus, for multi-indices $\alpha \in \{0, 1, \ldots, M-1\}^N$ we let

$$\{e_\alpha : |\alpha| \leq M-1\}, \qquad e_\alpha \in C_0^\infty(B(0,a)),$$

be a system dual to the monomials $\{y^\beta : |\beta| \leq M-1\}$, i.e.,

$$\int_{\mathbf{R}^N} e_\alpha y^\alpha \, dy = 1, \qquad \int_{\mathbf{R}^N} e_\alpha y^\beta \, dy = 0 \quad \text{for } \alpha \neq \beta.$$

We can assume that

$$\max_y |D^\gamma e_\alpha(y)| \leq C a^{-(|\alpha|+|\gamma|+N)}$$

for all $|\alpha| \leq M-1$ and $|\gamma| \leq L$. We set

$$\pi_f = \sum_{|\alpha| \leq M-1} \langle f, e_\alpha \rangle y^\alpha,$$

and for a $\psi \in C_0^\infty(B(0,a))$ such that $\|\psi\|_{C^L} \leq 1$, we set

$$\tilde{\psi} = \psi - \sum_{|\alpha| \leq M-1} \langle \psi, y^\alpha \rangle e_\alpha.$$

Then $\tilde{\psi} \in C_0^\infty(B(0,a))$, $\tilde{\psi} \perp \mathfrak{P}_{M-1}$, and

$$\langle f - \pi_f, \psi \rangle = \langle f, \psi \rangle - \sum_{|\alpha| \leq M-1} \langle f, e_\alpha \rangle \langle y^\alpha, \psi \rangle = \langle f, \tilde{\psi} \rangle.$$

We observe that $\|\psi\|_{L_\infty} \leq Ca^L$, and thus $|\langle \psi, y^\alpha \rangle| \leq Ca^{L+|\alpha|+N}$. It follows that for $|\gamma| = L$,

$$|D^\gamma \tilde{\psi}(y)| \leq |D^\gamma \psi(y)| + C \sum_{|\alpha| \leq M-1} a^{L+|\alpha|+N} |D^\gamma e_\alpha(y)|$$

$$\leq 1 + C \sum_{|\alpha| \leq M-1} a^{|\alpha|+L+N} a^{-(|\alpha|+L+N)} \leq C.$$

Thus $\|\tilde{\psi}\|_{C^L} \leq C$, and the lemma follows. $\square$

The following is our main result.

THEOREM 3.2.2. *Let* $\varepsilon_+, \varepsilon_- \in \mathbf{R}$, $r > 0$, *let* L *be a positive integer such that* $L > N \max\{\frac{1}{r} - 1, 0\} - \varepsilon_+$, *and let* $E \in S(\varepsilon_+, \varepsilon_-, r)$.

(a) *If* $s \in \mathbf{N}_0$, $0 \leq s \leq \varepsilon_-$, *then for any* $f \in Y(E)$ *and any* $\varepsilon > 0$ *there is an open set* G *with* $\mathrm{cap}(G, s, E) < \varepsilon$ *and a function* $g \in W_\infty^s \cap Y(E)$ *such that* g *agrees with* f *on* G^c *in the sense that for any* $x \in G^c$

(3.2.3) $$\sup_{0<a\leq 1} a^{-(s+N+L)} \mathcal{T}_0(f - g, B(x,a), L) \leq C\varepsilon^{-1} \|f\|_{Y(E)},$$

and

(3.2.4) $$\lim_{a \to 0} a^{-(s+N+L)} \mathcal{T}_0(f - g, B(x,a), L) = 0.$$

Moreover,

(3.2.5) $$\|g\|_{W_\infty^s} \leq C\varepsilon^{-1} \|f\|_{Y(E)},$$

(3.2.6) $$\|g\|_{Y(E)} \leq C\|f\|_{Y(E)},$$

and g *has a differential of order* s *everywhere in* $\mathbf{R}^N$ *in the sense that for each* x *there is a polynomial* $\pi \in \mathfrak{P}_s$ *such that*

(3.2.7) $$\lim_{a \to 0} a^{-(s+N+L)} \mathcal{T}_0(g - \pi, B(x,a), L) = 0.$$

If, in addition, $C^s \cap Y(E)$ *is dense in* $Y(E)$, *then one can also require that* $g \in C^s$. *In this case* $\|f - g\|_{Y(E)}$ *can be made arbitrarily small.*

(b) *If* $s \in \mathbf{R}$, $-\infty < s \leq \varepsilon_-$, *then for any* $f \in Y(E)$ *and any* $\varepsilon > 0$ *there is an open set* G *with* $\Lambda(G, s, E) < \varepsilon$ *and a function or distribution* $g \in B_{\infty,\infty}^s \cap Y(E)$ *such that* (3.2.3) *is satisfied for all* $x \in G^c$. *Moreover,*

(3.2.8) $$\|g\|_{B_{\infty,\infty}^s} \leq C\varepsilon^{-1} \|f\|_{Y(E)},$$

(3.2.9) $$\|g\|_{Y(E)} \leq C\|f\|_{Y(E)}.$$

If, in addition, $B_{\infty,\infty,0}^s \cap Y(E)$ *is dense in* $Y(E)$, *then one can also require that* $g \in B_{\infty,\infty,0}^s$. *In this case* (3.2.4) *holds for all* $x \in G^c$, *and* $\|f - g\|_{Y(E)}$ *can be made arbitrarily small.*

The proof of the theorem depends on the following Propositions 3.2.3 and 3.2.5.

PROPOSITION 3.2.3. *Let* $f \in \mathcal{S}'$, *let* M *and* L *be nonnegative integers such that* $L \leq M$, *and let* $-\infty < s < M$. *Define a set* G *by*

(3.2.10) $$G = \left\{ x \in \mathbf{R}^N : \mathcal{T}_0(f, B(x,1), L) + \sup_{0<a\leq 1} \frac{\mathcal{T}_M(f, B(x,a), L)}{a^{s+N+L}} > \lambda \right\}.$$

Then there is a function or distribution $g \in B_{\infty,\infty}^s(\mathbf{R}^N)$ *with* $\|g\|_{B_{\infty,\infty}^s} \leq C\lambda$ *which has the following properties:*

(i) For any $x \in G^c$ and any a, $0 < a \le 1$, we have

(3.2.11) $$\frac{\mathcal{T}_0(f-g, B(x,a), L)}{a^{s+N+L}} \le C\lambda;$$

(ii) There are constants C_1 and C_2 such that

(3.2.12) $$\mathcal{T}_M(g, B(x,a), L) \le C_1 \mathcal{T}_M(f, B(x, C_2 a), L)$$

for all x and a, $0 < a \le 1$, such that $B(x, 4a) \cap G^c \ne \emptyset$;

(iii) For all x and a, $0 < a \le 1$, such that $B(x, 4a) \subset G$

(3.2.13) $\mathcal{T}_M(g, B(x,a), L)$
$$\le C\Big(\frac{a}{\operatorname{dist}(x, \partial G)}\Big)^{M+N+L} \mathcal{T}_M(f, B(x, \tfrac{1}{2}\operatorname{dist}(x, \partial G)), L).$$

The proof of the proposition depends on the following lemma.

LEMMA 3.2.4. *Let $f \in \mathcal{S}'$, let $0 < R_1 < R_2 \le 1$, and let $M \in \mathbf{N}_0$. Let $\varphi \in C_0^\infty(B(0, R_2))$ be such that $0 \le \varphi \le 1$, and $\varphi = 1$ on $B(0, R_1)$. Then there exists a unique polynomial $\pi = \sum_{|\alpha| \le M-1} a_\alpha x^\alpha \in \mathfrak{P}_{M-1}$ such that $\varphi(f - \pi) \perp \mathfrak{P}_{M-1}$.*

If, moreover, $f \in L_\infty(B(0, R_2))$, then there is a C, depending only on R_2/R_1, N, and M, such that

(3.2.14) $$|a_\alpha| \le C\|f\|_{L_\infty(B(0,R_2))}, \quad |\alpha| \le M-1.$$

PROOF. We denote by c_M the cardinality of the set of multi-indices $\{\alpha : 0 \le |\alpha| \le M-1\}$. If we define a symmetric matrix $D = (d_{\alpha, \beta})$ of order c_M by setting

$$d_{\alpha, \beta} = \int_{B(0, R_2)} x^\alpha x^\beta \varphi(x)\, dx, \quad |\alpha|, |\beta| \le M-1,$$

the problem is to find a vector $(a_\alpha)_{|\alpha| \le M-1} = (a_\alpha) \in \mathbf{R}^{c_M}$, which satisfies the linear system

(3.2.15) $$\sum_{|\alpha| \le M-1} d_{\alpha, \beta} a_\alpha = \langle \varphi f, x^\beta \rangle, \quad |\beta| \le M-1.$$

Clearly, for any vector $(y_\alpha) \in \mathbf{R}^{c_M}$ the matrix product $(y_\alpha)^T D(y_\alpha)$ satisfies

$$(y_\alpha)^T D(y_\alpha) = \int_{B(0,R_2)} \Big(\sum_{|\alpha| \le M-1} y_\alpha x^\alpha\Big)^2 \varphi(x)\, dx$$
$$\ge \int_{B(0,R_1)} \Big(\sum_{|\alpha| \le M-1} y_\alpha x^\alpha\Big)^2 dx.$$

It follows that there is a constant $C_1 > 0$ such that

$$(y_\alpha)^T D(y_\alpha) \ge C_1 R_1^N \sum_{|\alpha| \le M-1} y_\alpha^2.$$

Thus all eigenvalues λ_α of D satisfy $\lambda_\alpha \ge C_1 R_1^N$, and consequently the determinant $\det D$ satisfies $\det D \ge (C_1 R_1^N)^{c_M} > 0$. Thus, the system (3.2.15) has a unique solution (a_α) for any f.

If $f \in L_\infty(B(0, R_2))$, then for all β

$$|\langle \varphi f, x^\beta \rangle| = \Big|\int_{B(0,R_2)} \varphi(x) f(x) x^\beta\, dx\Big| \le C_2 R_2^N \|f\|_{L_\infty(B(0,R_2))}.$$

Since $|d_{\alpha,\beta}| \leq C_3 R_2^N$, it follows easily from Cramer's rule that

$$|a_\alpha| \leq C_4 (C_1 R_1^N)^{-c_M} R_2^N \|f\|_{L_\infty(B(0,R_2))} (C_3 R_2^N)^{c_M-1}$$
$$= C(R_2/R_1)^{c_M} \|f\|_{L_\infty(B(0,R_2))},$$

where C only depends on N and M. □

PROOF OF PROPOSITION 3.2.3. The proof is analogous to that of Proposition 3.1.3, and we use the same notation. We again define g by setting

(3.2.16) $$g = f + \sum_{Q \in \mathcal{Q}} \varphi_Q (\pi_Q - f),$$

where the difference is that we now have to choose the polynomials π_Q in a more intricate way, and that à priori g is now only a distribution in $\mathcal{S}'$.

In order to prove (3.2.11) and (3.2.12) we fix a point x, and a, $0 < a \leq 1$, so that $B(x, 4a) \cap G^c \neq \emptyset$. We need to estimate $\sum_{Q \in \mathcal{Q}} \langle \varphi_Q (\pi_Q - f), \psi \rangle$, where $\psi \in C_0^\infty(B(x, a))$, and $\|\psi\|_{C^L} \leq 1$.

Here, for each Q, using Lemma 3.2.11, we choose π_Q as the unique polynomial in $\mathcal{P}_{M-1}$, which has the property that $\varphi_Q(\pi_Q - f) \perp \mathfrak{P}_{M-1}$, i.e.,

$$\langle \varphi_Q(\pi_Q - f), p \rangle = 0 \quad \text{for all } p \in \mathfrak{P}_{M-1}.$$

It follows that for any $p \in \mathfrak{P}_{M-1}$

$$\langle \varphi_Q(f - \pi_Q), \psi \rangle = \langle \varphi_Q(f - \pi_Q), \psi - p \rangle = \langle f - \pi_Q, \varphi_Q(\psi - p) \rangle.$$

Applying Lemma 3.2.4 once more, we choose p as the unique polynomial $p_Q \in \mathfrak{P}_{M-1}$ such that $\varphi_Q(\psi - p_Q) \perp \mathfrak{P}_{M-1}$. Setting $\varphi_Q(\psi - p_Q) = \tilde{\psi}_Q$ it follows that

$$\langle f - \pi_Q, \varphi_Q(\psi - p_Q) \rangle = \langle f - \pi_Q, \tilde{\psi}_Q \rangle = \langle f, \tilde{\psi}_Q \rangle.$$

In order to estimate $\|\tilde{\psi}_Q\|_{C^L}$ we observe that we can obtain p_Q in two steps, by first choosing $p'_Q \in \mathfrak{P}_{M-1}$ so that $\|D^\alpha(\psi - p'_Q)\|_{L_\infty(Q^*)} < C|Q|^{(L-|\alpha|)/N}$ for all $|\alpha| \leq L$, which is possible by the assumption $L \leq M$, and then setting $p_Q = p'_Q + p''_Q$, where p''_Q is the polynomial in $\mathfrak{P}_{M-1}$ satisfying $\varphi_Q(\psi - p'_Q - p''_Q) \perp \mathfrak{P}_{M-1}$, which is obtained by applying Lemma 3.2.4 to $\psi - p'_Q$.

It follows from (3.2.14) that $\|D^\alpha p''_Q\|_{L_\infty(Q^*)} < C|Q|^{(L-|\alpha|)/N}$ for all $|\alpha| \leq L$. By the assumption (3.1.12) we have $\|D^\beta \varphi_Q\|_{L_\infty} < C|Q|^{(-|\beta|)/N}$ for $|\beta| \leq L$, and by the Leibniz rule we obtain

$$\|\tilde{\psi}_Q\|_{C^L} \leq C \|\varphi_Q(\psi - p'_Q)\|_{C^L} \leq C.$$

Now, setting $\sum_{Q \in \mathcal{Q}} \tilde{\psi}_Q = \tilde{\psi}$, we have

$$\sum_{Q \in \mathcal{Q}} \langle \varphi_Q(\pi_Q - f), \psi \rangle = \sum_{Q \in \mathcal{Q}} \langle f, \tilde{\psi}_Q \rangle = \langle f, \tilde{\psi} \rangle.$$

But because of the bounded number of intersections of the Whitney cubes

$$\|\sum_{Q \in \mathcal{Q}} \tilde{\psi}_Q\|_{C^L} \leq C \max_{Q \in \mathcal{Q}} \|\tilde{\psi}_Q\|_{C^L} \leq C,$$

and thus, $\tilde{\psi} \in C_0^\infty(B(x, C_2 a))$, $\|\tilde{\psi}\|_{C^L} \leq C$, and $\tilde{\psi} \perp \mathfrak{P}_{M-1}$.

If $x \in G^c$ it follows that

$$|\langle f - g, \psi \rangle| = |\sum_{Q \in \mathcal{Q}} \langle \varphi_Q(\pi_Q - f), \psi \rangle| = |\langle f, \tilde{\psi} \rangle| \leq C \lambda a^{s+N+L},$$

which proves (3.2.11).

3.2. LUZIN APPROXIMATION OF DISTRIBUTIONS

If in addition it is assumed that ψ satisfies $\psi \perp \mathfrak{P}_{M-1}$, and if $B(x,4a) \cap G^c \neq \emptyset$, we obtain

$$|\langle g, \psi \rangle| \leq |\langle f, \psi \rangle| + \left|\sum_{Q \in \mathfrak{Q}} \langle \varphi_Q(\pi_Q - f), \psi \rangle\right|$$
$$= |\langle f, \psi \rangle| + |\langle f, \tilde{\psi} \rangle| \leq C \mathcal{T}_M(f, B(x, C_2 a), L),$$

which proves (3.2.12).

In order to prove (3.2.13) we let $0 < a \leq 1$, and we assume that $x \in G$ and that $\mathrm{dist}(x, \partial G) > 4a$. Using Lemma 3.2.1 we choose $\pi \in \mathfrak{P}_{M-1}$ so that

$$(3.2.17) \quad \mathcal{T}_0(f - \pi, B(x, \tfrac{1}{2}\mathrm{dist}(x, \partial G)), L) \leq C \mathcal{T}_M(f, B(x, \tfrac{1}{2}\mathrm{dist}(x, \partial G)), L).$$

Then $g - \pi = \sum_{Q \in \mathfrak{Q}} \varphi_Q(\pi_Q - \pi)$ in $B(x, a)$. For a $\psi \in C_0^\infty(B(x, a))$ such that $\psi \perp \mathfrak{P}_{M-1}$ we obtain

$$\langle g, \psi \rangle = \langle g - \pi, \psi \rangle = \sum_{Q \in \mathfrak{Q}} \langle \varphi_Q(\pi_Q - \pi), \psi \rangle$$
$$= \sum_{Q \in \mathfrak{Q}} \langle \varphi_Q(\pi_Q - \pi) - P_x^{M-1}(\varphi_Q(\pi_Q - \pi)), \psi \rangle,$$

where P_x^{M-1} again denotes the Taylor polynomial of degree $M-1$ at x. As in the proof of Proposition 3.1.3 it is enough to estimate one of the terms such that Q^* intersects $B(x, a)$. We find

$$|\langle \varphi_Q(\pi_Q - \pi) - P_x^{M-1}(\varphi_Q(\pi_Q - \pi)), \psi \rangle|$$
$$\leq \int_{B(x,a)} |\varphi_Q(\pi_Q - \pi) - P_x^{M-1}(\varphi_Q(\pi_Q - \pi))| |\psi| \, dy$$
$$\leq C a^{M+N} \max_{|\gamma|=M} \|D^\gamma(\varphi_Q(\pi_Q - \pi))\|_{L_\infty} \|\psi\|_{L_\infty}.$$

We need to estimate $\|\pi_Q - \pi\|_{L_\infty(Q^*)}$. As before there is by Lemma 1.2.2 a constant C such that $\|\pi_Q - \pi\|_{L_\infty(Q^*)} \leq C|Q|^{-1} \|\pi_Q - \pi\|_{L_1(Q)}$, and since $\mathcal{T}_0(\cdot, Q, L)$ is a norm on $\mathfrak{P}_{M-1}$, there is also a constant C, independent of $|Q|$ and L, such that

$$\|\pi_Q - \pi\|_{L_1(Q)} \leq C \sup_{\psi \in C_0^\infty(Q)} \frac{1}{|Q|^{L/N} \|\psi\|_{C^L}} \left|\int_Q (\pi_Q - \pi)\psi \, dy\right|$$
$$\leq C|Q|^{-L/N} \mathcal{T}_0(\pi_Q - \pi, Q, L).$$

But

$$\mathcal{T}_0(\pi_Q - \pi, Q, L) \leq \mathcal{T}_0(f - \pi, Q, L) + \mathcal{T}_0(f - \pi_Q, Q, L),$$

and by the assumption (3.1.11) and (3.2.17)

$$\mathcal{T}_0(f - \pi, Q, L) \leq C \mathcal{T}_0(f - \pi, B(x, \tfrac{1}{2}\mathrm{dist}(x, \partial G)), L)$$
$$\leq C \mathcal{T}_M(f, B(x, \tfrac{1}{2}\mathrm{dist}(x, \partial G)), L).$$

On the other hand, for a suitable $\psi_0 \in C_0^\infty(Q)$ with $\|\psi_0\|_{C^L} \leq 1$, we have by Lemma 3.2.4 as in the proof of Proposition 3.2.3,

$$\tfrac{1}{2}\mathcal{T}_0(f - \pi_Q, Q, L) \leq \langle f - \pi_Q, \psi_0 \rangle$$
$$= \langle \varphi_Q(f - \pi_Q), \psi_0 \rangle = \langle f - \pi_Q, \varphi_Q \psi_0 \rangle = \langle f, \tilde{\psi}_0 \rangle,$$

where $\tilde{\psi}_0 \in C_0^\infty(Q^*)$, $\tilde{\psi}_0 \perp \mathfrak{P}_{M-1}$, and $\|\tilde{\psi}_0\|_{C^L} \leq C$. It follows that

$$\mathcal{T}_0(f - \pi_Q, Q, L) \leq C \mathcal{T}_M(f, Q^*, L) \leq C \mathcal{T}_M(f, B(x, \tfrac{1}{2}\mathrm{dist}(x, \partial G)), L).$$

The Leibniz rule gives as before that
$$\max_{|\gamma|=M} \|D^\gamma(\varphi_Q(\pi_Q - \pi))\|_{L_\infty} \leq \frac{C\mathcal{T}_M(f, B(x, \tfrac{1}{2}\operatorname{dist}(x, \partial G)), L)}{\operatorname{dist}(x, \partial G)^{M+N+L}}.$$

Thus
$$|\langle g, \psi\rangle| \leq Ca^{M+N}\|\psi\|_{L_\infty} \frac{C\mathcal{T}_M(f, B(x, \tfrac{1}{2}\operatorname{dist}(x, \partial G)), L)}{\operatorname{dist}(x, \partial G)^{M+N+L}},$$

which, combined with the fact that $\|\psi\|_{L_\infty} \leq Ca^L$, gives the desired inequality (3.2.13).

Finally, it follows as in the proof of Proposition 3.1.3 that
$$\sup_{0 < a \leq 1} a^{-(s+N+L)} \mathcal{T}_M(g, B(x, a), L) \leq C\lambda$$

for all x. Thus, by e.g. Theorem 1.1.15(ii), g belongs to $B^s_{\infty,\infty}$, and $\|g\|_{B^s_{\infty,\infty}} \leq C\lambda$. In particular g is a function if $s > 0$. $\square$

PROPOSITION 3.2.5. *Let $f \in \mathcal{S}'$, let L and M be nonnegative integers such that $L \leq M$, and let s be an integer such that $0 \leq s \leq M - 1$. Let $G \subset \mathbf{R}^N$ be the set of points x such that*

(3.2.18) $$\mathcal{T}_0(f, B(x, 1), L) + \sum_{i=1}^\infty 2^{i(s+N+L)} \mathcal{T}_M(f, B(x, 2^{-i}), L) > \lambda.$$

Then there exists a function $g \in W^s_\infty$ with $\|g\|_{W^s_\infty} < C\lambda$ such that (i) – (iii) in Proposition 3.2.3 hold true, and such that for all $x \in G^c$ we have

(3.2.19) $$\sup_{0 < a \leq 1} a^{-(s+N+L)} \mathcal{T}_0(f - g, B(x, a), L) \leq C\lambda,$$

and

(3.2.20) $$\lim_{a \to 0} a^{-(s+N+L)} \mathcal{T}_0(f - g, B(x, a), L) = 0.$$

Moreover, g has a differential of order s everywhere in $\mathbf{R}^N$ in the sense that for each x there is a polynomial $\pi \in \mathfrak{P}_s$ such that

(3.2.21) $$\lim_{a \to 0} a^{-(s+N+L)} \mathcal{T}_0(g - \pi, B(x, a), L) = 0.$$

PROOF. Let G be an open set of points such that (3.2.18) is satisfied, and define g by (3.2.16) as before. It follows in the same way as (3.1.16) follows from (3.1.17) that there is a C such that

(3.2.22) $$\mathcal{T}_0(g, B(x, 1), L) + \sum_{i=1}^\infty 2^{i(s+N+L)} \mathcal{T}_M(g, B(x, 2^{-i}), L) \leq C\lambda$$

for all x. The relations (3.2.19) and (3.2.20) follow immediately.

By Lemma 3.2.1 there are polynomials $\pi_{i,k} \in \mathfrak{P}_{M-1}$ such that
$$\mathcal{T}_0(g - \pi_{i,k}, Q_{i,k}(3), L) \leq C\mathcal{T}_M(g, Q_{i,k}(3), L), \quad (i, k) \in \mathbf{N}_0 \times \mathbf{Z}^N.$$

Since $\mathcal{T}_0(\,\cdot\,, Q_{i,k}(3), L)$ is a norm on $\mathfrak{P}_{M-1}$, there is as in Lemma 1.2.2 a constant C, independent of i, such that if $Q_{i,k}(3)$ and $Q_{i-1,l}(3)$ intersect,

$$\|\pi_{i,k} - \pi_{i-1,l}\|_{L_\infty(Q_{i,k}(3))} \leq C 2^{i(N+L)} \mathcal{T}_0(\pi_{i,k} - \pi_{i-1,l}, Q_{i,k}(3), L)$$
$$\leq C 2^{i(N+L)} (\mathcal{T}_0(g - \pi_{i,k}, Q_{i,k}(3), L) + \mathcal{T}_0(g - \pi_{i-1,l}, Q_{i-1,l}(3), L))$$
$$\leq C 2^{i(N+L)} \mathcal{T}_0(g - \pi_{i-1,l}, Q_{i-1,l}(3), L).$$

As in the proof of Proposition 3.1.4 we define functions $g_i(x) = \sum_{k \in \mathbf{Z}^N} s_{i,k} a_{i,k}(x)$, so that g has an atomic representation $g(x) = \sum_{i=i_0}^{\infty} g_i(x)$, where i_0 can chosen so large that for each x

$$|s_{i,k} a_{i,k}(x)| \leq |s_{i,k}| \leq C 2^{i(N+L)} \mathcal{T}_M(g, B(x, 2^{-i+i_0}), L).$$

It follows that

$$\sum_{i=i_0}^{\infty} |D^\beta g_i(x)| \leq \sum_{i=i_0}^{\infty} 2^{i|\beta|} 2^{i(N+L)} \mathcal{T}_M(g, B(x, 2^{-i+i_0}), L)$$

for all $|\beta| \leq s$, which by (3.2.22) implies that $g \in W_s^\infty$ with $\|g\|_{W_s^\infty} \leq C\lambda$.

It remains to prove (3.2.7). The proof of this is essentially the same as the last part of the proof of Proposition 3.1.4. We fix without loss of generality $x = 0$, and we denote

$$2^{i(N+L)} \mathcal{T}_M(g, B(0, 2^{-i}), L) = T_i,$$

so that

(3.2.23) $$\sum_{i=0}^{\infty} 2^{is} T_i \leq C\lambda.$$

For each $i \in \mathbf{N}_0$ we define $\pi_i \in \mathfrak{P}_{M-1}$ by setting $\pi_0 \equiv T_0$, and by choosing π_i for $i \geq 1$ so that

$$\mathcal{T}_0(g - \pi_i, B(0, 2^{-i}), L) \leq C \mathcal{T}_M(g, B(0, 2^{-i}), L),$$

which is possible by Lemma 3.2.1. Then

$$\|\pi_i - \pi_{i+1}\|_{L_\infty(B(0, 2^{-i-1}))} \leq C 2^{i(N+L)} \mathcal{T}_M(g, B(0, 2^{-i}), L) = C T_i,$$

for $i \geq 1$. If we write $\pi_i(y) = \sum_{|\alpha| \leq M-1} a_{i,\alpha} y^\alpha$, it follows from the equivalence of norms on finite dimensional spaces that

(3.2.24) $$|a_{i,\alpha} - a_{i+1,\alpha}| \leq C 2^{i|\alpha|} T_i$$

for $|\alpha| \leq M - 1$. For $|\alpha| \leq s$ it follows from (3.2.24) that

(3.2.25) $$\sum_{i=0}^{\infty} |a_{i,\alpha} - a_{i+1,\alpha}| \leq C\lambda,$$

and that

$$\lim_{i \to \infty} a_{i,\alpha} = a_{0,\alpha} + \sum_{i=0}^{\infty} (a_{i+1,\alpha} - a_{i,\alpha}) = a_\alpha$$

exists with

(3.2.26) $$|a_\alpha| \leq C\lambda.$$

For $s + 1 \leq |\alpha| \leq M - 1$ the inequality (3.2.24) gives similarly

(3.2.27) $$|a_{j,\alpha}| \leq C \sum_{i=0}^{j} 2^{i|\alpha|} T_i.$$

We define $\pi(y) = \sum_{|\alpha| \leq s} a_\alpha y^\alpha$, and we claim that (3.2.24) and (3.2.27) imply

(3.2.28) $$\lim_{j \to \infty} 2^{j(s+N+L)} \mathcal{T}_0(g - \pi, B(0, 2^{-j}), L) = 0,$$

as required. We write for each j

$$g(y) - \pi(y) = (g(y) - \pi_j(y))$$
$$+ \sum_{|\alpha|\leq s}(a_{j,\alpha} - a_\alpha)y^\alpha + \sum_{s+1\leq|\alpha|\leq M-1} a_{j,\alpha}y^\alpha = G_1 + G_2 + G_3,$$

and we estimate the three terms separately. By the choice of π_j we already have

$$\lim_{j\to\infty} 2^{j(s+N+L)}\mathcal{T}_0(g - \pi_j, B(0, 2^{-j}), L) \leq C \lim_{j\to\infty} 2^{js}T_j = 0.$$

In estimating the second and third terms we observe that

$$\langle G_2, \psi\rangle = \int_{B(0,2^{-j})} G_2(y)\psi(y), dy$$

and similarly for G_3. If $\psi \in C_0^\infty(B(0,2^{-j}))$ and $\|\psi\|_{C^L} \leq 1$, then $\|\psi\|_{L^\infty} \leq C2^{-jL}$. By (3.2.24)

$$|a_{j,\alpha} - a_\alpha| \leq \sum_{i=j}^\infty |a_{i,\alpha} - a_{i+1,\alpha}| \leq C\sum_{i=j}^\infty 2^{i|\alpha|}T_i,$$

and thus for all such ψ

$$2^{j(N+L)}|\langle G_2, \psi\rangle| \leq 2^{jN}\int_{B(0,2^{-j})}\sum_{|\alpha|\leq s}|(a_{j,\alpha} - a_\alpha)y^\alpha|\,dy$$
$$\leq C\sum_{|\alpha|\leq s}2^{-j|\alpha|}\sum_{i=j}^\infty 2^{i|\alpha|}T_i \leq 2^{-js}\sum_{i=j}^\infty 2^{is}T_i,$$

which by (3.2.23) gives the desired estimate. In the third term (3.2.27) gives

$$2^{j(N+L)}|\langle G_3, \psi\rangle| \leq 2^{jN}\int_{B(0,2^{-j})}\sum_{s+1\leq|\alpha|\leq M-1}|a_{j,\alpha}y^\alpha|\,dy$$
$$\leq C\sum_{s+1\leq|\alpha|\leq M-1}2^{-j|\alpha|}\sum_{i=0}^j 2^{i|\alpha|}T_i \leq C2^{-js}\sum_{i=0}^j 2^{i-j}2^{is}T_i.$$

The required estimate again follows from the convergence of $\sum_{i=0}^\infty 2^{is}T_i$, and the claim (3.2.28) is proved. $\square$

PROOF OF THEOREM 3.2.2. Let $f \in Y(E)$, and let $M \geq [\varepsilon_-] + 1$ and $M \geq L$. By Theorem 1.1.15(ii), the functions

$$h_0(x) = \mathcal{T}_0(g, B(x,1), L), \quad h_i(x) = 2^{i(N+L)}\mathcal{T}_M(g, B(x, 2^{-i}), L), \quad i = 1, 2, \ldots,$$

satisfy $\|\{h_i\}_{i=0}^\infty\|_E \leq C\|f\|_{Y(E)}$. Thus, for any s and any $\lambda > 0$, the set

$$G_s = \left\{y : \sum_{i=0}^\infty 2^{is}h_i(y) > \lambda\right\}$$

satisfies

$$\mathrm{cap}(G_s, s, E) < C\lambda^{-1}\|f\|_{Y(E)}.$$

Similarly, the set

$$H_s = \left\{y : \sup_{i\in\mathbf{N}_0} 2^{is}h_i(y) > \lambda\right\}$$

satisfies

$$\Lambda(H_s, s, E) < C\lambda^{-1}\|f\|_{Y(E)}.$$

It follows from Proposition 3.2.3 that if $\varepsilon > 0$ is given, and if λ is chosen so that $\lambda = C\varepsilon^{-1}\|f\|_{YL(E)}$, then the function $g \in B^s_{\infty,\infty}$ constructed in the proposition satisfies (3.2.3) and (3.2.8), and it agrees with f in the sense of (3.2.3) for all x off a set H_s such that $\Lambda(H_s, s, E) < \varepsilon$. Moreover, g satisfies (3.2.12) and (3.2.13), and this easily implies (3.2.9) by Theorem 1.1.15(ii).

If s is an integer, $s = 0, 1, \ldots, M - 1$, it follows in the same way from Proposition 3.2.5 that there is a function $g \in W^s_\infty$, which agrees with f in the sense of (3.2.3) and (3.2.4) for all x off a set G_s such that $\mathrm{cap}(G_s, s, E) < \varepsilon$, and which also satisfies the conditions (3.2.5), (3.2.6), and (3.2.7).

If we finally assume that smooth functions are dense in $Y(E)$, the conclusions follow almost exactly as in the proof of Theorem 3.1.2, and the argument need not be repeated. $\square$

and, since clearly $\Omega_M(f,t,L_p(Q)) \leq C\omega_M(f,t,L_p(Q))$, we have as an immediate consequence

(A.5) $$\mathcal{E}_M(f,Q,L_p) \leq C\omega_M(f,\rho a, L_p(Q)).$$

The latter is actually the form in which the Whitney theorem is usually stated.

The function $\omega_k(f,t,L_p(A))$ has the advantage that it is increasing as a function of t, whereas $\Omega_k(f,t,L_p(A))$ can more easily be added for different sets.

We shall need both properties, so it will be important that the two modules are essentially equivalent, as shown by the following lemma. As before we denote the cube, concentric to Q with λ times the side of Q, by $Q(\lambda)$.

LEMMA A.2. *There is a constant $C = C(k,p)$ such that for any cube Q with side a and every $\lambda > 1$*

(A.6) $$\omega_k(f,t,L_p(Q)) \leq C\Omega_k(f,(1+k)t,L_p(Q(\lambda)))$$

for $0 < t \leq a(\lambda - 1)/2k$.

PROOF. We follow [15], Ch. 6, Lemma 5.1. We shall use the following identity, which holds for any z and w in $\mathbf{R}^N$.

(A.7) $$\Delta_z^k f(x) = \sum_{i=1}^k (-1)^i \binom{k}{i} \left(\Delta_{iw}^k f(x+iz) - \Delta_{z+iw}^k f(x) \right).$$

To prove it we use the definition of Δ_z^k and a change of order of summation to obtain

$$\sum_{i=0}^k (-1)^{k-i} \binom{k}{i} \Delta_{z+iw}^k f(x)$$
$$= \sum_{i=0}^k \sum_{j=0}^k (-1)^{k-i} \binom{k}{i} (-1)^{k-j} \binom{k}{j} f(x+jz+ijw)$$
$$= \sum_{j=0}^k (-1)^{k-j} \binom{k}{j} \Delta_{jw}^k f(x+jz).$$

It is now enough to observe that the term corresponding to $j=0$ is zero.

From (A.7) we get (with a C depending on p in the case $p > 1$)

$$\|\Delta_z^k f\|_{L_p(Q)}^p \leq C \sum_{i=1}^k \left(\|\Delta_{iw}^k f\|_{L_p(Q+iz)}^p + \|\Delta_{z+iw}^k f\|_{L_p(Q)}^p \right).$$

Taking the average over $|w| < t$, assuming that $|z| < t \leq a$, and noting that then $|z+iw| < (1+k)t$ and $Q + iz \subset Q(1+2k(t/a))$, we obtain after changing variables

$$\|\Delta_z^k f\|_{L_p(Q)}^p \leq Ct^{-N} \int_{|w|<(1+k)t} \|\Delta_w^k f\|_{L_p(Q(1+2k(t/a)))}^p \, dw,$$

whence

$$\omega_k(f,t,L_p(Q))^p \leq Ct^{-N} \int_{|w|<(1+k)t} \|\Delta_w^k f\|_{L_p(Q(1+2k(t/a)))}^p \, dw$$
$$\leq C\Omega_k(f,(1+k)t,L_p(Q(\lambda)))^p, \quad \text{if } t \leq a(\lambda-1)/2k.$$

$\square$

As a first step towards the proof of Theorem A.1 we prove the following special case.

PROPOSITION A.3. *Let $f \in L_{p,\text{loc}}(\mathbf{R}^N)$, $0 < p \leq \infty$. If $\Delta_z^M f(x) = 0$ for almost all $(x, z) \in R^{2N}$, then f is almost everywhere equal to a polynomial in $\mathfrak{P}_{M-1}$.*

PROOF. By the formula $\Delta_z^M = (T_z - I)^M$ we have

$$f = (-1)^M \Delta_z^M f - \sum_{k=1}^{M} (-1)^k \binom{M}{k} T_{kz} f. \tag{A.8}$$

First let $p \geq 1$. If $\varphi \in C_0^\infty$ and $\int_{\mathbf{R}^N} \varphi(z)\,dz = 1$, it follows from the assumptions and from Fubini's theorem that for almost all x

$$f(x) = f(x) \int_{\mathbf{R}^N} \varphi(z)\,dz = \sum_{k=1}^{M} (-1)^{k+1} \binom{M}{k} \int_{\mathbf{R}^N} \varphi(z) f(x + kz)\,dz$$

$$= \sum_{k=1}^{M} (-1)^{k+1} \binom{M}{k} k^{-N} \int_{\mathbf{R}^N} \varphi((z-x)/k) f(z)\,dz.$$

But each of the terms on the right hand side belongs to C^∞, so $f \in C^\infty$ after redefinition on a set of measure zero. Thus, for all directions ξ the directional derivatives $D_\xi^M \equiv 0$, which easily implies that $D^\alpha f \equiv 0$ for all $|\alpha| = M$, and thus $f \in \mathfrak{P}_{M-1}$. (Note that a similar convolution argument gives the same conclusion if f is a distribution such that $\Delta_z^M f = 0$ for almost all z. One then uses the theorem of L. Schwartz [46] Ch. II, Th. VI (Cor.), saying that if $f \in \mathcal{D}$ satisfies $D^\alpha f = 0$ for all $|\alpha| = M$, then $f \in \mathfrak{P}_{M-1}$.)

Now let $0 < p < 1$. We claim that the assumptions imply that f is locally bounded. By (A.8) we have for almost all x, and z

$$|f(x)|^p \leq C \sum_{k=1}^{M} |f(x + kz)|^p.$$

It follows that

$$|f(x)|^p \leq C \sum_{k=1}^{M} \int_{|z|<1} |f(x + kz)|^p\,dz,$$

and here the right hand side is a locally bounded function of x. It follows from the first part of the proof that $f \in \mathfrak{P}_{M-1}$. $\square$

REMARK. Proposition A.3 has, of course, a long history. Already Cauchy proved that the only continuous solutions of the functional equation $f(x + z) = f(x) + f(z)$ on $\mathbf{R}$ are linear, and in 1905 G. Hamel [23] described by means of the Zermelo axiom the discontinuous solutions of the same equation, and proved that they are all unbounded. In 1908 L. E. J. Brouwer [9] proved a theorem on uniform differentiability which immediately gives Proposition A.3 for continuous functions of N variables. (We are grateful to Yu. A. Brudnyi for this and several other references.) Brouwer's result implies, in fact, that $f \in \mathfrak{P}_{M-1}$ if $\lim_{z \to 0} \Delta_z^M f(x) |z|^{-M} = 0$ uniformly in x. Apparently unaware of [9], H. Whitney [56], obtained this conclusion for (Lebesgue) measurable functions f on $\mathbf{R}$. That bounded functions satisfying $\Delta_z^M f(x) \equiv 0$ belong to $\mathfrak{P}_{M-1}$ without the assumption of measurability is a special case of the result in [59]. A different proof was given by B. Sendov and V. A. Popov [47], Th. 2.1.

Brudnyĭ [10] (Theorem 1), and [11] (Lemma 6) proved Proposition A.3 in the case $p \geq 1$, but his proofs do not extend to $0 < p < 1$. Nevskiĭ [38], Lemma 7, stated the general result for $0 < p < \infty$ but gave no proof. A proof for $0 < p < \infty$ in one dimension, which easily extends to the N-dimensional case, is given by DeVore and Lorentz [15], Ch. 12, Th. 5.3. Their proof is by induction, and depends on a highly non-obvious identity of J. H. B. Kemperman. (See Johnen and Scherer [28], Lemma 2.) We do not know any reference for the proof given above.

The proof of Theorem A.1 consists in combining Proposition A.3 with the following condition for precompactness, which goes back to M. Riesz [43].

PROPOSITION A.4. *Let $0 < p < \infty$, let Q be a cube, and $0 < \lambda < 1$. A subset $\mathcal{G}$ of $L_p(Q)$ is precompact in $L_p(Q(\lambda))$ if there is a constant C such that $\|f\|_{L_p(Q)} \leq C$ for all $f \in \mathcal{G}$, and if*

(A.9) $$\lim_{t \to 0} \Omega_M(f, t, L_p(Q)) = 0, \quad \textit{uniformly for } f \in \mathcal{G}.$$

The proof of the proposition can be reduced to the case $M = 1$ (the case treated by Riesz) by means of the following inequality, which is a slight modification of a classical inequality of A. Marchaud [29]. We formulate and prove the inequality for the case $0 < p \leq 1$, and omit the easy changes needed if $p > 1$. We again essentially follow [15], Section 2.8.

LEMMA A.5. *Let $0 < p \leq 1$, and let $M \in \mathbf{N}$. Then for every $\lambda > 1$ there are positive constants C and t_0, depending on M, p, and λ, such that for any cube Q with side a, and $k = 1, 2, \ldots, M - 1$,*

(A.10) $$\Omega_k(f, t, L_p(Q))^p \leq C t^{kp} \left(\frac{\|f\|^p_{L_p(Q(\lambda))}}{a^{kp}} + \int_t^a \frac{\Omega_M(f, s, L_p(Q(\lambda))^p}{s^{kp+1}} \, ds \right)$$

for $0 < t < t_0$.

PROOF. We first prove the inequality for $M = k+1$, $k = 1, 2, 3, \ldots$. Define a polynomial $P \in \mathfrak{P}_{k-1}$ by

$$P(x) = \frac{1 - 2^{-k}(x+1)^k}{x - 1},$$

so that

$$(x-1)^k = 2^{-k}(x^2 - 1)^k + P(x)(x-1)^{k+1}.$$

Replacing x by the translation operator T_z we obtain

$$(T_z - 1)^k = 2^{-k}(T_{2z} - 1)^k + P(T_z)(T_z - 1)^{k+1},$$

which gives the identity

$$\Delta_z^k f(x) = 2^{-k} \Delta_{2z}^k f(x) + P(T_z) \Delta_z^{k+1} f(x),$$

due to Marchaud ([29], p. 48). Each T_z^j is an operator of norm 1 on L_p, and it follows that there is a constant C_P, not exceeding the sum of the absolute values of the coefficients of P, such that

$$\|P(T_z) \Delta_z^{k+1} f\|_{L_p(Q)} \leq C_P \|\Delta_z^{k+1} f\|_{L_p(Q(1 + 2(k-1)|z|/a))}.$$

Thus

$$\|\Delta_z^k f\|^p_{L_p(Q)} \leq 2^{-kp} \|\Delta_{2z}^k f\|^p_{L_p(Q)} + C_P^p \|\Delta_z^{k+1} f\|^p_{L_p(Q(1+2(k-1)|z|/a))}.$$

Iterating m times, and assuming that $2^m k|z| \leq (\lambda - 1)a$, we find

$$\|\Delta_z^k f\|_{L_p(Q)}^p \leq 2^{-mkp}\|\Delta_{2^m z}^k f\|_{L_p(Q)}^p + C_Q^p \sum_{j=0}^{m-1} 2^{-jkp}\|\Delta_{2^j z}^{k+1} f\|_{L_p(Q(1+2^{j+1}k|z|/a))}^p$$

$$\leq C 2^{-mkp}\|f\|_{L_p(Q(\lambda))}^p + C_Q^p \sum_{j=0}^{m-1} 2^{-jkp}\|\Delta_{2^j z}^{k+1} f\|_{L_p(Q(\lambda))}^p.$$

Now let t satisfy $0 < 2kt < (\lambda - 1)a$. We determine $m \in \mathbf{N}$ so that $2^m kt < (\lambda - 1)a \leq 2^{m+1}kt$. Averaging over $|z| \leq t$ gives

$$\Omega_k(f, t, L_p(Q))^p \leq C\left(2^{-mkp}\|f\|_{L_p(Q(\lambda))}^p + \sum_{j=0}^{m-1} 2^{-jkp}\Omega_{k+1}(f, 2^j t, L_p(Q(\lambda)))^p\right).$$

But clearly

$$\Omega_{k+1}(f, t, L_p(Q(\lambda)))^p \leq 2^N \Omega_{k+1}(f, 2t, L_p(Q(\lambda)))^p,$$

whence

$$\Omega_{k+1}(f, 2^j t, L_p(Q(\lambda)))^p \leq C 2^{jkp} t^{kp} \int_{2^j t}^{2^{j+1} t} \frac{\Omega_{k+1}(f, s, L_p(Q(\lambda)))^p}{s^{kp+1}} ds.$$

By the choice of m we obtain

$$\Omega_k(f, t, L_p(Q))^p$$

(A.11)
$$\leq C t^{kp}\left(\frac{\|f\|_{L_p(Q(\lambda))}^p}{((\lambda-1)a)^{kp}} + \sum_{j=0}^{m-1} \int_{2^j t}^{2^{j+1} t} \frac{\Omega_{k+1}(f, s, L_p(Q(\lambda)))^p}{s^{kp+1}} ds\right)$$

$$\leq C t^{kp}\left(\frac{\|f\|_{L_p(Q(\lambda))}^p}{((\lambda-1)a)^{kp}} + \int_t^{(\lambda-1)a/k} \frac{\Omega_{k+1}(f, s, L_p(Q(\lambda)))^p}{s^{kp+1}} ds\right)$$

for $0 < t \leq (\lambda-1)a/2k$.

In the general case the proposition now follows by induction on M. We assume that for $M = k + l$

(A.12) $\quad \Omega_k(f, t, L_p(Q)^p) \leq C t^{kp}\left(\frac{\|f\|_{L_p(Q(\lambda^l))}^p}{a^{kp}} + \int_t^a \frac{\Omega_M(f, s, L_p(Q(\lambda^l))^p}{s^{kp+1}} ds\right),$

and we claim that (A.12) holds for k and $M = k + l + 1$. By (A.11)

$$\int_t^a \frac{\Omega_{k+l}(f, s, L_p(Q(\lambda^l))^p}{s^{kp+1}} ds$$

$$\leq C \frac{\|f\|_{L_p(Q(\lambda^{l+1}))}^p}{a^{(k+l)p}} \int_t^a s^{lp-1} ds$$

$$+ C \int_t^a s^{lp-1}\left(\int_s^a \frac{\Omega_{k+l+1}(f, u, L_p(Q(\lambda^{l+1}))^p}{u^{kp+lp+1}} du\right) ds$$

$$\leq C \frac{\|f\|_{L_p(Q(\lambda^{l+1}))}^p}{a^{kp}} + C \int_t^a \frac{\Omega_{k+l+1}(f, u, L_p(Q(\lambda^{l+1}))^p}{u^{kp+1}} du,$$

where the last inequality follows from a change of order of integration. The induction is finished by substituting this in (A.12). The desired inequality (A.10) now follows from (A.12) if λ is replaced by $\lambda^{1/l}$. □

PROOF OF PROPOSITION A.4. Let $\mathcal{G} \subset L_p(Q)$ satisfy the conditions of the proposition. By Lemma A.5 it follows that

(A.13) $$\lim_{t \to 0} \Omega_1(f, t, L_p(Q(\lambda))) = 0, \quad \text{uniformly for } f \in \mathcal{G}.$$

We can assume that $Q(\lambda)$ is the unit cube $Q_{0,0}$. For each $i = 1, 2, \ldots$, we subdivide the unit cube into 2^{iN} cubes $Q_{i,k}$ with side 2^{-i}. We shall approximate the functions in $\mathcal{G}$ by piecewise constant functions g_i.

For each $Q_{i,k}$ there is a constant $c_{i,k}$ such that

(A.14) $$2^{iN} \int_{Q_{i,k}} \int_{Q_{i,k}} |f(y) - f(x)|^p \, dx \, dy = \int_{Q_{i,k}} |f(x) - c_{i,k}|^p \, dx.$$

In fact, $\int_{Q_{i,k}} |f(y) - f(x)|^p \, dx$ is a continuous function of y, and therefore by the mean value theorem there is an $\eta_{i,k} \in Q_{i,k}$ such that $c_{i,k} = f(\eta_{i,k})$ satisfies (A.14).

We define g_i by
$$g_i(x) = \sum_k c_{i,k} \chi_{i,k}(x),$$

where the summation is taken over such k that $Q_{i,k} \subset Q_{0,0}$. Then (A.14) gives

$$\int_{Q_{0,0}} |f(x) - g_i(x)|^p \, dx = \sum_k \int_{Q_{i,k}} |f(x) - c_{i,k}|^p \, dx$$
$$= 2^{iN} \sum_k \int_{Q_{i,k}} \int_{Q_{i,k}} |f(y) - f(x)|^p \, dx \, dy$$
$$\leq 2^{iN} \int_{|z| < 2^{-i+1}\sqrt{N}} \int_{Q_{0,0}} |f(x+z) - f(x)|^p \, dx \, dz$$
$$\leq C \Omega_1(f, 2^{-i+1}\sqrt{N}, L_p(Q_{0,0}))^p,$$

and consequently by (A.13), for an arbitrarily given $\varepsilon > 0$,

(A.15) $$\|f - g_i\|_{L_p(Q_{0,0})} < \varepsilon,$$

if i is large enough.

By assumption the set $\mathcal{G}_i$ of all such step functions g_i corresponding to functions $f \in \mathcal{G}$ is bounded in $L_p(Q_{0,0})$. Since $\mathcal{G}_i$ is a subset of a finite dimensional subspace of $L_p(Q_{0,0})$ it has a finite ε-net. It follows from (A.15) that $\mathcal{G}$ has a finite $C_p \varepsilon$-net for a constant C_p only depending on p. In other words, $\mathcal{G}$ is totally bounded, and thus precompact. (See e.g. DiBenedetto [16], I.17.) $\square$

REMARK. With a slightly greater effort Proposition A.4 can be proved with $\lambda = 1$. The idea is to write the difference $f(x+z) - f(x)$ as a sum of N differences along the coordinate directions, and then apply Marchaud's inequality to each such coordinate in the corresponding two halves of the cube. See Nevskiĭ [38].

PROOF OF THEOREM A.1. It is enough to prove the theorem for $a = 1$. We claim that for $0 < \rho \leq 1$ there is a constant C such that

(A.16) $$\mathcal{E}_M(f, Q, L_p) \leq C \Omega_M(f, \rho, L_p(Q)).$$

Suppose the contrary. Then for each $n = 1, 2, \ldots$, there exists an $f_n \in L_p(Q)$ with $\|f_n\|_{L_p(Q)} = 1$, $\mathcal{E}_M(f_n, Q, L_p) \geq \frac{1}{2}$, and $\Omega_M(f_n, \rho, L_p(Q)) \leq \frac{1}{n}$.

Let $0 < \lambda < 1$. Then $Q' = Q(\lambda)$ is a cube contained in the interior of Q. By Lemma A.2 there are t' and C, depending on λ but not on n, such that
$$\omega_M(f_n, t, L_p(Q')) \leq C\Omega_M(f_n, (1+M)t, L_p(Q)) \quad \text{for } 0 < t \leq t'.$$
Thus, by Lemma A.2 and the monotonicity of ω, for $0 < t \leq t'$,
$$\Omega_M(f_n, t, L_p(Q')) \leq C\omega_M(f_n, t, L_p(Q'))$$
$$\leq C\omega_M(f_n, t', L_p(Q')) \leq C\Omega_M(f_n, (1+M)t', L_p(Q)).$$
But, if t' is chosen so that $(1+M)t' \leq \rho$, obviously
$$\Omega_M(f_n, (1+M)t', L_p(Q)) \leq (\rho/(1+M)t')^{N/p}\Omega_M(f_n, \rho, L_p(Q)) \leq Cn^{-1}.$$
If $\varepsilon > 0$ is given, and we choose n_0 sufficiently large, it follows that
$$\Omega_M(f_n, t, L_p(Q')) < \varepsilon$$
for $0 < t \leq t'$ and $n \geq n_0$.

Also, $\Omega_M(f_n, t, L_p(Q'))$ is continuous, and $\lim_{t \to 0} \Omega_M(f_n, t, L_p(Q')) = 0$, so there is a $t'' > 0$ such that
$$\Omega_M(f_n, t, L_p(Q')) < \varepsilon \quad \text{for } 0 < t \leq t'' \text{ for } n = 1, 2, \ldots, n_0.$$
Thus, $\Omega_M(f_n, t, L_p(Q')) < \varepsilon$ for all $0 < t \leq \min\{t', t''\}$ for all $n \in \mathbf{N}$. By Proposition A.4 this implies that $\mathcal{G} = \{f_n\}_{n=0}^{\infty}$ is precompact in $L_p(Q'')$ for any $Q'' = \lambda''Q$, where $0 < \lambda'' < \lambda'$. It follows that we can extract a subsequence which converges in $L_p(Q')$ to a function f. But then $\Omega_M(f, \rho, L_p(Q'')) = 0$, so that $\Delta_z^M f(x) = 0$ almost everywhere, and thus, by Proposition A.3, f is a polynomial in $\mathfrak{P}_{M-1}$ on Q''. But λ'' can be chosen arbitrarily close to 1, so f is a polynomial in $\mathfrak{P}_{M-1}$ on Q. This contradicts the fact that $\mathcal{E}_M(f, Q, L_p) = \lim_{n \to \infty} \mathcal{E}_M(f_n, Q, L_p) \geq \frac{1}{2}$, and the contradiction proves the theorem. □

Bibliography

[1] Adams, D. R., A note on Choquet integrals with respect to Hausdorff capacity, *Function Spaces and Applications* (Proc. Conf. Lund 1986), *Lecture Notes in Math.* **1302**, 115–124, Springer, Berlin Heidelberg, 1988.

[2] _____ The classification problem for the capacities associated with the Besov and Triebel–Lizorkin spaces, *Approximation and Function Spaces* (Proc. Warsaw, 1986), *Banach Center Publ.* **22**, 9–24, PWN, Warsaw, 1989.

[3] _____ Choquet integrals in potential theory, *Publ. Mat.* **42** (1998), 3–66.

[4] Adams, D. R. and Hedberg, L. I., *Function Spaces and Potential Theory*, Springer, Berlin Heidelberg, 1996.

[5] Andersen, K. F and John, R. T., Weighted inequalities for vector-valued maximal functions and singular integrals, *Studia Math.* **69** (1980/81), no. 1, 19–31.

[6] Aoki, Tosio, Locally bounded linear topological spaces, *Proc. Imp. Acad. Tokyo* **18** (1942), 588–594.

[7] Bergh, J. and Löfström, J., *Interpolation Spaces. An Introduction*, Springer, Berlin Heidelberg, 1976.

[8] Bojarski, B., Hajłasz, P. and Strzelecki, P., Improved $C^{k,\lambda}$ approximation of higher order Sobolev functions in norm and capacity, *Indiana Univ. Math. J.* **51** (2002), no. 3, 507–540.

[9] Brouwer, L. E. J., About difference quotients and differential quotients, *Proc. Sec. Sci, Koninklijke Nederlandse Akademie van Wetenschappen* **11** (1908), 59–66. Reprinted in *L. E. J. Brouwer, Collected Works, Vol. 2*, North-Holland, Amsterdam, 1976, pp. 93–101.

[10] Brudnyĭ, Yu. A. (Брудный, Ю. А.), Критерии существования производных в L^p, *Mat. Sb.* **73(115)** (1967), no. 1, 42–64. English translation: Criteria for the existence of derivatives in L^p, *Math. USSR-Sb.* **2** (1967), 35–55.

[11] _____ Многомерный аналог одной теоремы Уитни, *Mat. Sb.* **82(124)** (1970), no. 2(6), 175–191. English translation: A multidimensional analogue of a certain theorem of Whitney, *Math. USSR-Sb.* **11** (1970), 157–170.

[12] _____ Пространства, определяемые с помощью локальных приближений, *Trudy Moskov. Mat. Obshch.* **24** (1971), 69–132. English translation: Spaces defined by means of local approximations, *Trans. Moscow Math. Soc.*, 1971, 73–139 (1974).

[13] _____ Неравенство Уитни для квазибанаховых пространтв (Whitney's inequality for quasi-Banach spaces), *Функциональные пространства и их применение к дифференциальным уравнениям* (*Function spaces and their application to differential equations*), 20–27, Ed. V. Maslennikova, Izd. Ross. univ. druzhby narodov, Moscow, 1992.

[14] _____ Адаптивная аппроксимация функций с особенностями, *Trudy Moskov. Mat. Obshch.* **55** (1994), 149–242. English translation: Adaptive approximation of functions with singularities, *Trans. Moscow Math. Soc.*, 1994, 123–186 (1995).

[15] DeVore, R. A. and Lorentz, G. G., *Constructive Approximation*, Springer, Berlin Heidelberg, 1993.

[16] DiBenedetto, E., *Real Analysis*, Birkhäuser, Boston, 2002.

[17] Fefferman, C. and Stein, E. M., Some maximal inequalities, *Amer. J. Math.* **93** (1971), 107–115.

[18] _____ H^p-spaces of several variables, *Acta Math.* **129** (1972), 137–193.

[19] Frazier, M. and Jawerth, B., Decomposition of Besov spaces, *Indiana Univ. Math. J.* **34** (1985), 777–799.

[20] _____ The φ-transform and applications to distribution spaces, *Function Spaces and Applications*, Proc., Lund 1986 (M. Cwikel, J. Peetre, Y. Sagher, H. Wallin, eds.), Lecture Notes in Math. **1302**, 223–246, Springer, Berlin Heidelberg, 1988.

[21] _____ A discrete transform and decompositions of distribution spaces, *J. Funct. Anal.* **93** (1990), 34–170.

[22] Frazier, M., Jawerth, B. and Weiss, G., *Littlewood–Paley Theory and the Study of Function Spaces*, CBMS Regional Conf. Ser. in Math. **79**, Amer. Math. Soc., Providence, Rhode Island, 1991.

[23] Hamel, G., Eine Basis aller Zahlen und die unstetigen Lösungen der Funktionalgleichung: $f(x+y) = f(x) + f(y)$, *Math. Ann.* **60** (1905), 459–462.

[24] Hedberg, L. I., Spectral synthesis in Sobolev spaces, and uniqueness of solutions of the Dirichlet problem, *Acta Math.* **147** (1981), 237–264.

[25] Heinonen, J., *Lectures on Analysis on Metric Spaces*, Springer, New York, 2001.

[26] Hörmander, L., *The Analysis of Linear Partial Differential Operators I*, Springer, Berlin Heidelberg, 1983.

[27] Jawerth, B., Pérez, C. and Welland, G., The positive cone in Triebel–Lizorkin spaces and the relation among potential and maximal operators, *Harmonic Analysis and Partial Differential Equations*, Proc. Conf., Boca Raton, 1988 (M. Milman and T. Schonbek, eds.), 71–91, Contemporary Mathematics **107**, Amer. Math. Soc., Providence, Rhode Island, 1989.

[28] Johnen, H. and Scherer, K., On the equivalence of the K-functional and moduli of continuity and some applications, *Constructive Theory of Functions of Several Variables*, Proc. Conf., Math. Res. Inst., Oberwolfach, 1976, 119–140, Lecture Notes in Math. **571**, Springer, Berlin, 1977.

[29] Marchaud, A., Sur les dérivées et sur les differences des fonctions de variables réelles, *J. Math. pures appl.*, **6** (1927), 337–425.

[30] Netrusov, Yu. V. (Нетрусов, Ю. В.), Теоремы вложения пространств Бесова в идеальные пространства, *Zap. Nauchn. Sem. Leningrad. Otdel. Mat. Inst. Steklov. (LOMI)* **159** (1987), 69–82. English translation: Imbedding theorems of Besov spaces in Banach lattices, *J. Soviet Math.* **47**:6 (1989), 2871–2881.

[31] _____ Теоремы вложения пространств Лизоркина–Трибеля, *Zap. Nauchn. Sem. Leningrad. Otdel. Mat. Inst. Steklov. (LOMI)* **159** (1987), 103–112. English translation: Embedding theorems in Lizorkin–Triebel spaces, *J. Soviet Math.* **47**:6 (1989), 2896–2903.

[32] _____ Множества особенностей функций из пространств Бесова (Exceptional sets of functions from Besov spaces), *Vsesoyuznaya shkola po teorii funkciĭ (12–22 oktyabrya 1987), Tezisy dokladov*, Izdatel'stvo Erevanskogo universiteta, Erevan, 1987.

[33] _____ Множества особенностей функций из пространств типа Бесова и Лизоркина–Трибеля, *Trudy Matem. Inst. Steklov.* **187** (1989), 162–177. English translation: Sets of singularities of functions in spaces of Besov and Lizorkin–Triebel type, *Proc. Steklov Inst. Math.* **187** (1990), 185–203.

[34] _____ Метрические оценки емкостей множеств в пространствах Бесова, *Trudy Matem. Inst. Steklov.* **190** (1989), 159–185. English translation: Metric estimates of the capacities of sets in Besov spaces, *Proc. Steklov Inst. Math.* **190** (1992), 167–192.

[35] _____ Спектральный синтез в пространствах гладких функций, *Ross. Akad. Nauk Dokl.* **325** (1992), 923–925. English translation: Spectral synthesis in spaces of smooth functions, *Russian Acad. Sci. Dokl. Math.* **46** (1993), 135–137.

[36] _____ Оценки емкостей, связанных с пространствами Бесова, *Zap. Nauchn. Sem. S.-Peterburg. Otdel. Mat. Inst. Steklov. (POMI)* **201** (1992), 124–156. English translation: Estimates of capacities associated with Besov spaces, *J. Math. Sci.* **78** no. 2 (1996), 199–217.

[37] _____ Спектральный синтез в пространстве Соболева, порожденном интегральной метрикой, *Zap. Nauchn. Sem. S.-Peterburg. Otdel. Mat. Inst. Steklov. (POMI)* **217** (1994), 217–234. English translation: Spectral synthesis in a Sobolev space generated by an integral metric, *J. Math. Sci. (New York)* **85** no. 2 (1997), 1814–1826.

[38] Nevskiĭ, M. V. (Невский, М. В.), О приближении функций в классах Орлича (Approximation of functions in Orlicz classes), *Исследования по теории функций многих вещественных переменных* (*Studies in the theory of functions of several real variables*), 83–101, Ed. Yu. A. Brudnyĭ, Yaroslav. Gos. Univ., Yaroslavl', 1984.

[39] Peetre, J., Sur les espaces de Besov, *C. R. Acad. Sci. Paris Sér. A-B* **264** (1967), A281–A283.

[40] _____ Remarques sur les espaces de Besov. Le cas $0 < p < 1$, *C. R. Acad. Sci. Paris Sér. A-B* **277** (1973), A947–A949.

[41] _____ On spaces of Triebel–Lizorkin type, *Ark. mat.* **13** (1975), 123–130. Correction, *ibid.* **14** (1976), 299.

[42] _____ *New Thoughts on Besov Spaces* Mathematics Department, Duke University, Durham, North Carolina, 1976.

[43] Riesz, M., Sur les ensembles compacts de fonctions sommables, *Acta Litt. Sci. Szeged* **6** (1933), 136-142. Reprinted in *Riesz, Marcel: Collected Papers*, 458–464, Springer, Berlin, 1988.

[44] Rolewicz, S., On a certain class of linear metric spaces. *Bull. Acad. Polon. Sci. Cl. III.* **5** (1957), 471–473.

[45] _____ *Metric Linear Spaces*, 2nd ed., PWN–Polish Scientific Publishers, Warsaw; D. Reidel Publishing Co., Dordrecht, 1984.

[46] Schwartz, L., *Théorie des distributions*, 2nd ed., Hermann, Paris, 1966.

[47] Sendov, B. and Popov, V. A. (Сендов, Б., Попов, В. А.), *Усреднени модули на гладкост* (in Bulgarian), Publ. House Bulg. Ac. Sci., Sofia, 1983. English translation: *The Averaged Moduli of Smoothness*, Wiley, Chichester, 1988. Russian translation: *Усредненные модули гладкости*, Mir, Moscow, 1988.

[48] Stein, E. M., *Singular Integrals and Differentiability of Functions*, Princeton University Press, Princeton, New Jersey, 1970.

[49] _____ *Harmonic Analysis: Real-Variable Methods, Orthogonality, and Oscillatory Integrals*, Princeton University Press, Princeton, New Jersey, 1993.

[50] Storozhenko, È. A. (Стороженко, Э. А.), О приближении алгебраическими многочленами функций класса L^p, $0 < p < 1$ (Approximation by algebraic polynomials of functions of the class L^p, $0 < p < 1$), *Izv. Akad. Nauk SSSR, Ser. Mat.* **41** no. 3 (1977), 652–662.

[51] Storozhenko, È. A. and Osval′d, P. [Oswald, P.] (Стороженко, Э. А., Освальд, П.), Теорема Джексона в пространствах $L^p(R^k)$, $0 < p < 1$, *Sibirsk. Mat. Zh.* **19** no. 4 (1978), 888–901. English translation: Jackson's theorem in the spaces $L^p(R^k)$, $0 < p < 1$, *Siberian Math. J.*, **19** no. 4 (1978), 630–640 (1979).

[52] Swanson, D., Pointwise inequalities and approximation in fractional Sobolev spaces, *Studia Math.* **149** (2002), 147–174.

[53] Triebel, H., *Theory of Function Spaces*, Akademische Verlagsgesellschaft Geest & Portig K.-G., Leipzig, 1983, and Birkhäuser Verlag, Basel, 1983.

[54] _____ *Theory of Function Spaces II*, Birkhäuser Verlag, Basel, 1992.

[55] _____ Spaces on sets, *Uspekhi Mat. Nauk*, S. M. Nikolskii centennial issue, to appear.

[56] Whitney, H., Derivatives, difference quotients, and Taylor's formula, *Bull. Amer. Math. Soc.* **40** (1934), 89–94.

[57] _____ On ideals of differentiable functions, *Amer. J. Math.* **70** (1948), 635–658.

[58] _____ On functions with bounded nth differences, *J. Math. Pures Appl.* (9) **36** (1957), 67–95.

[59] _____ On bounded functions with bounded nth differences, *Proc. Amer. Math. Soc.*, **10** (1959), 480–481.

[60] Ziemer, W. P., *Weakly Differentiable Functions*, Springer, New York, 1989.

Editorial Information

To be published in the *Memoirs*, a paper must be correct, new, nontrivial, and significant. Further, it must be well written and of interest to a substantial number of mathematicians. Piecemeal results, such as an inconclusive step toward an unproved major theorem or a minor variation on a known result, are in general not acceptable for publication.

Papers appearing in *Memoirs* are generally at least 80 and not more than 200 published pages in length. Papers less than 80 or more than 200 published pages require the approval of the Managing Editor of the Transactions/Memoirs Editorial Board.

As of February 28, 2007, the backlog for this journal was approximately 15 volumes. This estimate is the result of dividing the number of manuscripts for this journal in the Providence office that have not yet gone to the printer on the above date by the average number of monographs per volume over the previous twelve months, reduced by the number of volumes published in four months (the time necessary for preparing a volume for the printer). (There are 6 volumes per year, each usually containing at least 4 numbers.)

A Consent to Publish and Copyright Agreement is required before a paper will be published in the *Memoirs*. After a paper is accepted for publication, the Providence office will send a Consent to Publish and Copyright Agreement to all authors of the paper. By submitting a paper to the *Memoirs*, authors certify that the results have not been submitted to nor are they under consideration for publication by another journal, conference proceedings, or similar publication.

Information for Authors

Memoirs are printed from camera copy fully prepared by the author. This means that the finished book will look exactly like the copy submitted.

Initial submission. The AMS uses Centralized Manuscript Processing for initial submissions. Authors should submit a PDF file using the Initial Manuscript Submission form found at www.ams.org/cgi-bin/peertrack/submission.pl, or send one copy of the manuscript to the following address: Centralized Manuscript Processing, MEMOIRS OF THE AMS, 201 Charles Street, Providence, RI 02904-2294 USA. If a paper copy is being forwarded to the AMS, indicate that it is for it Memoirs and include the name of the corresponding author, contact information such as email address or mailing address, and the name of an appropriate Editor to review the paper (see the list of Editors below).

The paper must contain a *descriptive title* and an *abstract* that summarizes the article in language suitable for workers in the general field (algebra, analysis, etc.). The *descriptive title* should be short, but informative; useless or vague phrases such as "some remarks about" or "concerning" should be avoided. The *abstract* should be at least one complete sentence, and at most 300 words. Included with the footnotes to the paper should be the 2000 *Mathematics Subject Classification* representing the primary and secondary subjects of the article. The classifications are accessible from www.ams.org/msc/. The list of classifications is also available in print starting with the 1999 annual index of *Mathematical Reviews*. The Mathematics Subject Classification footnote may be followed by a list of *key words and phrases* describing the subject matter of the article and taken from it. Journal abbreviations used in bibliographies are listed in the latest *Mathematical Reviews* annual index. The series abbreviations are also accessible from www.ams.org/publications/. To help in preparing and verifying references, the AMS offers MR Lookup, a Reference Tool for Linking, at www.ams.org/mrlookup/.

Electronically prepared manuscripts. The AMS encourages electronically prepared manuscripts, with a strong preference for $\mathcal{AMS}$-LaTeX. To this end, the Society has prepared $\mathcal{AMS}$-LaTeX author packages for each AMS publication. Author packages include instructions for preparing electronic manuscripts, samples, and a style file that generates the particular design specifications of that publication series. Though $\mathcal{AMS}$-LaTeX is the highly preferred format of TeX, author packages are also available in $\mathcal{AMS}$-TeX.

Authors may retrieve an author package from the AMS website starting from `www.ams.org/tex/` or via FTP to `ftp.ams.org` (login as `anonymous`, enter username as password, and type `cd pub/author-info`). The *AMS Author Handbook* and the *Instruction Manual* are available in PDF format following the author packages link from `www.ams.org/tex/`. The author package can also be obtained free of charge by sending email to `tech-support@ams.org` (Internet) or from the Publication Division, American Mathematical Society, 201 Charles St., Providence, RI 02904-2294, USA. When requesting an author package, please specify $\mathcal{AMS}$-LaTeX or $\mathcal{AMS}$-TeX and the publication in which your paper will appear. Please be sure to include your complete mailing address.

After acceptance. The final version of the electronic file should be sent to the Providence office (this includes any TeX source file, any graphics files, and the DVI or PostScript file) immediately after the paper has been accepted for publication.

Before sending the source file, be sure you have proofread your paper carefully. The files you send must be the EXACT files used to generate the proof copy that was accepted for publication. For all publications, authors are required to send a printed copy of their paper, which exactly matches the copy approved for publication, along with any graphics that will appear in the paper.

Accepted electronically prepared files can be submitted via the web at `www.ams.org/submit-book-journal/`, sent via FTP, or sent on CD-Rom or diskette to the Electronic Prepress Department, American Mathematical Society, 201 Charles Street, Providence, RI 02904-2294 USA. TeX source files, DVI files, and PostScript files can be transferred over the Internet by FTP to the Internet node `ftp.ams.org` (130.44.1.100). When sending a manuscript electronically via CD-Rom or diskette, please be sure to include a message identifying the paper as a Memoir.

Electronically prepared manuscripts can also be sent via email to `pub-submit@ams.org` (Internet). In order to send files via email, they must be encoded properly. (DVI files are binary and PostScript files tend to be very large.)

Electronic graphics. Comprehensive instructions on preparing graphics are available at `www.ams.org/jourhtml/`. A few of the major requirements are given here.

Submit files for graphics as EPS (Encapsulated PostScript) files. This includes graphics originated via a graphics application as well as scanned photographs or other computer-generated images. If this is not possible, TIFF files are acceptable as long as they can be opened in Adobe Photoshop or Illustrator. No matter what method was used to produce the graphic, it is necessary to provide a paper copy to the AMS.

Authors using graphics packages for the creation of electronic art should also avoid the use of any lines thinner than 0.5 points in width. Many graphics packages allow the user to specify a "hairline" for a very thin line. Hairlines often look acceptable when proofed on a typical laser printer. However, when produced on a high-resolution laser imagesetter, hairlines become nearly invisible and will be lost entirely in the final printing process.

Screens should be set to values between 15% and 85%. Screens which fall outside of this range are too light or too dark to print correctly. Variations of screens within a graphic should be no less than 10%.

Inquiries. Any inquiries concerning a paper that has been accepted for publication should be sent to `memo-query@ams.org` or directly to the Electronic Prepress Department, American Mathematical Society, 201 Charles St., Providence, RI 02904-2294 USA.

Editors

This journal is designed particularly for long research papers, normally at least 80 pages in length, and groups of cognate papers in pure and applied mathematics. Papers intended for publication in the *Memoirs* should be addressed to one of the following editors. The AMS uses Centralized Manuscript Processing for initial submissions to AMS journals. Authors should follow instructions listed on the Initial Submission page found at www.ams.org/memo/memosubmit.html.

Algebra to ALEXANDER KLESHCHEV, Department of Mathematics, University of Oregon, Eugene, OR 97403-1222; email: ams@noether.uoregon.edu

Algebra and its application to MINA TEICHER, Emmy Noether Research Institute for Mathematics, Bar-Ilan University, Ramat-Gan 52900, Israel; email: teicher@macs.biu.ac.il

Algebraic geometry to DAN ABRAMOVICH, Department of Mathematics, Brown University, Box 1917, Providence, RI 02912; email: amsedit@math.brown.edu

Algebraic number theory to V. KUMAR MURTY, Department of Mathematics, University of Toronto, 100 St. George Street, Toronto, ON M5S 1A1, Canada; email: murty@math.toronto.edu

Algebraic topology to ALEJANDRO ADEM, Department of Mathematics, University of British Columbia, Room 121, 1984 Mathematics Road, Vancouver, British Columbia, Canada V6T 1Z2; email: adem@math.ubc.ca

Combinatorics to JOHN R. STEMBRIDGE, Department of Mathematics, University of Michigan, Ann Arbor, Michigan 48109-1109; email: FRS@umich.edu

Complex analysis and harmonic analysis to ALEXANDER NAGEL, Department of Mathematics, University of Wisconsin, 480 Lincoln Drive, Madison, WI 53706-1313; email: nagel@math.wisc.edu

Differential geometry and global analysis to LISA C. JEFFREY, Department of Mathematics, University of Toronto, 100 St. George St., Toronto, ON Canada M5S 3G3; email: jeffrey@math.toronto.edu

Dynamical systems and ergodic theory to AMIE WILKINSON, Department of Mathematics, Northwestern University, 2033 Sheridan Road, Evanston, IL 60208-2730; email: transactions@math.northwestern.edu

Functional analysis and operator algebras to DIMITRI SHLYAKHTENKO, Department of Mathematics, University of California, Los Angeles, CA 90095; email: shlyakht@math.ucla.edu

Geometric analysis to WILLIAM P. MINICOZZI II, Department of Mathematics, Johns Hopkins University, 3400 N. Charles St., Baltimore, MD 21218; email: trans@math.jhu.edu

Geometric analysis to MLADEN BESTVINA, Department of Mathematics, University of Utah, 155 South 1400 East, JWB 233, Salt Lake City, Utah 84112-0090; email: bestvina@math.utah.edu

Harmonic analysis, representation theory, and Lie theory to ROBERT J. STANTON, Department of Mathematics, The Ohio State University, 231 West 18th Avenue, Columbus, OH 43210-1174; email: stanton@math.ohio-state.edu

Logic to STEFFEN LEMPP, Department of Mathematics, University of Wisconsin, 480 Lincoln Drive, Madison, Wisconsin 53706-1388; email: lempp@math.wisc.edu

Partial differential equations to GUSTAVO PONCE, Department of Mathematics, South Hall, Room 6607, University of California, Santa Barbara, CA 93106; email: ponce@math.ucsb.edu

Partial differential equations and dynamical systems to PETER POLACIK, School of Mathematics, University of Minnesota, Minneapolis, MN 55455; email: polacik@math.umn.edu

Probability and statistics to KRZYSZTOF BURDZY, Department of Mathematics, University of Washington, Box 354350, Seattle, Washington 98195-4350; email: burdzy@math.washington.edu

Real analysis and partial differential equations to DANIEL TATARU, Department of Mathematics, University of California, Berkeley, Berkeley, CA 94720; email: tataru@math.berkeley.edu

All other communications to the editors should be addressed to the Managing Editor, ROBERT GURALNICK, Department of Mathematics, University of Southern California, Los Angeles, CA 90089-1113; email: guralnic@math.usc.edu.

Titles in This Series

883 **Apostolos Beligiannis and Idun Reiten,** Homological and homotopical aspects of torsion theories, 2007

882 **Lars Inge Hedberg and Yuri Netrusov,** An axiomatic approach to function spaces, spectral synthesis, and Luzin approximation, 2007

881 **Tao Mei,** Operator valued Hardy spaces, 2007

880 **Bruce C. Berndt, Geumlan Choi, Youn-Seo Choi, Heekyoung Hahn, Boon Pin Yeap, Ae Ja Yee, Hamza Yesilyurt, and Jinhee Yi,** Ramanujan's forty identities for the Rogers-Ramanujan functions, 2007

879 **O. García-Prada, P. B. Gothen, and V. Muñoz,** Betti numbers of the moduli space of rank 3 parabolic Higgs bundles, 2007

878 **Alessandra Celletti and Luigi Chierchia,** KAM stability and celestial mechanics, 2007

877 **María J. Carro, José A. Raposo, and Javier Soria,** Recent developments in the theory of Lorentz spaces and weighted inequalities, 2007

876 **Gabriel Debs and Jean Saint Raymond,** Borel liftings of Borel sets: Some decidable and undecidable statements, 2007

875 **C. Krattenthaler and T. Rivoal,** Hypergéométrie et fonction zêta de Riemann, 2007

874 **Sonia Natale,** Semisolvability of semisimple Hopf algebras of low dimension, 2007

873 **A. J. Duncan,** Exponential genus problems in one-relator products of groups, 2007

872 **Anthony V. Geramita, Tadahito Harima, Juan C. Migliore, and Yong Su Shin,** The Hilbert function of a level algebra, 2007

871 **Pascal Auscher,** On necessary and sufficient conditions for L^p-estimates of Riesz transforms associated to elliptic operators on $\mathbb{R}^n$ and related estimates, 2007

870 **Takuro Mochizuki,** Asymptotic behaviour of tame harmonic bundles and an application to pure twistor D-modules, Part 2, 2007

869 **Takuro Mochizuki,** Asymptotic behaviour of tame harmonic bundles and an application to pure twistor D-modules, Part 1, 2007

868 **Gelu Popescu,** Entropy and multivariable interpolation, 2006

867 **Vilmos Totik,** Metric properties of harmonic measures, 2006

866 **William Craig,** Semigroups underlying first-order logic, 2006

865 **Nathanial P. Brown,** Invariant means and finite representation theory of $C*$-algebras, 2006

864 **John M. Lee,** Fredholm operators and Einstein metrics on conformally compact manifolds, 2006

863 **M. Lübke and A. Teleman,** The Universal Kobayashi-Hitchin correspondence on Hermitian manifolds, 2006

862 **Alberto Canonaco,** The Beilinson complex and canonical rings of irregular surfaces, 2006

861 **Leon A. Takhtajan and Lee-Peng Teo,** Weil-Petersson metric on the universal Teichmüller space, 2006

860 **Thomas M. Fiore,** Pseudo limits, biadjoints and pseudo algebras: Categorical foundations of conformal field theory, 2006

859 **N. Arcozzi, R. Rochberg, and E. Sawyer,** Carleson measures and interpolating sequences for Besov spaces on complex balls, 2006

858 **Enrico Valdinoci, Berardino Sciunzi, and Vasile Ovidiu Savin,** Flat level set regularity of p-Laplace phase transitions, 2006

857 **Donatella Danielli, Nocola Garofalo, and Duy-Minh Nhieu,** Non-doubling Ahlfors measures, perimeter measures, and the characterization of the trace spaces of Sobolev functions in Carnot-Carathéodory spaces, 2006

856 **Vladimir Bolotnikov and Harry Dym,** On boundary interpolation for matrix valued Schur functions, 2006

TITLES IN THIS SERIES

855 **Yevgenia Kashina, Yorck Sommerhäuser, and Yongchang Zhu,** On higher Frobenius-Schur indicators, 2006

854 **Noam Greenberg,** The role of true finiteness in the admissible recursively enumerable degrees, 2006

853 **Joachim Krieger,** Stability of spherically symmetric wave maps, 2006

852 **Viorel Barbu, Irena Lasiecka, and Roberto Triggiani,** Tangential boundary stabilization of Navier-Stokes equations, 2006

851 **Jie Wu,** On maps from loop suspensions to loop spaces and the shuffle relations on the Cohen groups, 2006

850 **Siegfried Echterhoff, S. Kaliszewski, John Quigg, and Iain Raeburn,** A categorical approach to imprimitivity theorems for C^*-dynamical systems, 2006

849 **Katsuhiko Kuribayashi, Mamoru Mimura, and Tetsu Nishimoto,** Twisted tensor products related to the cohomology of the classifying spaces of loop groups, 2006

848 **Bob Oliver,** Equivalences of classifying spaces completed at the prime two, 2006

847 **Eric T. Sawyer and Richard L. Wheeden,** Hölder continuity of weak solutions to subelliptic equations with rough coefficients, 2006

846 **Victor Beresnevich, Detta Dickinson, and Sanju Velani,** Measure theoretic laws for lim–sup sets, 2006

845 **Ehud Friedgut, Vojtech Rödl, Andrzej Ruciński, and Prasad V. Tetali,** A Sharp threshold for random graphs with a monochromatic triangle in every edge coloring, 2006

844 **Amadeu Delshams, Rafael de la Llave, and Tere M. Seara,** A geometric mechanism for diffusion in Hamiltonian systems overcoming the large gap problem: Heuristics and rigorous verification on a model, 2006

843 **Denis V. Osin,** Relatively hyperbolic groups: Intrinsic geometry, algebraic properties, and algorithmic problems, 2006

842 **David P. Blecher and Vrej Zarikian,** The calculus of one-sided M-ideals and multipliers in operator spaces, 2006

841 **Enrique Artal Bartolo, Pierrette Cassou-Noguès, Ignacio Luengo, and Alejandro Melle Hernández,** Quasi-ordinary power series and their zeta functions, 2005

840 **Sławomir Kołodziej,** The complex Monge-Ampère equation and pluripotential theory, 2005

839 **Mihai Ciucu,** A random tiling model for two dimensional electrostatics, 2005

838 **V. Jurdjevic,** Integrable Hamiltonian systems on complex Lie groups, 2005

837 **Joseph A. Ball and Victor Vinnikov,** Lax-Phillips scattering and conservative linear systems: A Cuntz-algebra multidimensional setting, 2005

836 **H. G. Dales and A. T.-M. Lau,** The second duals of Beurling algebras, 2005

835 **Kiyoshi Igusa,** Higher complex torsion and the framing principle, 2005

834 **Kenı́chi Ohshika,** Kleinian groups which are limits of geometrically finite groups, 2005

833 **Greg Hjorth and Alexander S. Kechris,** Rigidity theorems for actions of product groups and countable Borel equivalence relations, 2005

832 **Lee Klingler and Lawrence S. Levy,** Representation type of commutative Noetherian rings III: Global wildness and tameness, 2005

831 **K. R. Goodearl and F. Wehrung,** The complete dimension theory of partially ordered systems with equivalence and orthogonality, 2005

For a complete list of titles in this series, visit the
AMS Bookstore at **www.ams.org/bookstore/**.